PHYSICAL OCEANOGRAPHIC PROCESSES OF THE GREAT BARRIER REEF

PHYSICAL OCEANOGRAPHIC PROCESSES OF THE GREAT BARRIER REEF

Eric Wolanski

With an Introduction by G. L. Pickard

CRC Press
Taylor & Francis Group
Boca Raton London New York

CRC Press is an imprint of the
Taylor & Francis Group, an **informa** business

First published 1994 by CRC Press
Taylor & Francis Group
6000 Broken Sound Parkway NW, Suite 300
Boca Raton, FL 33487-2742

Reissued 2018 by CRC Press

Library of Congress Cataloging-in-Publication Data

Wolanski, Eric.
Physical oceanographic processes of the Great Barrier Reef / Eric Wolanski ; with an introduction by G. L. Pickard.
p. cm. — (Marine science series)
Includes bibliographical references and index.
ISBN 0-8493-8047-2
1. Oceanography — Australia — Great Barrier Reef (Qld.) I. Title. II. Series.
GC862.W65 1994
551.46'576—dc20 93-31280

A Library of Congress record exists under LC control number: 93031280

Publisher's Note
The publisher has gone to great lengths to ensure the quality of this reprint but points out that some imperfections in the original copies may be apparent.

Disclaimer
The publisher has made every effort to trace copyright holders and welcomes correspondence from those they have been unable to contact.

ISBN 13: 978-1-315-89650-2 (hbk)
ISBN 13: 978-1-351-07560-2 (ebk)

Visit the Taylor & Francis Web site at http://www.taylorandfrancis.com and the
CRC Press Web site at http://www.crcpress.com

Marine Science Series

The CRC Marine Science Series provides publications that synthesize recent advances in Marine Science. Marine Science is at an exciting new threshold where new developments are providing fresh perspectives on how the biology of the ocean is integrated with its chemistry and physics.

CRC MARINE SCIENCE SERIES

SERIES EDITORS

Michael J. Kennish, Ph.D.
Peter L. Lutz, Ph.D.

PUBLISHED TITLES

Ecology of Estuaries: Anthropogenic Effects, Michael J. Kennish
The Physiology of Fishes, David H. Evans
Physical Oceanographic Processes of the Great Barrier Reef, Eric Wolanski

---

FORTHCOMING TITLES

Benthic Microbial Ecology, Paul F. Kemp
Chemosynthetic Communities, James M. Brooks and Chuck Fisher
Ecology of Marine Invertebrate Larvae, Larry McEdward
Effects of Coastal Pollution on Living Resources, Carl J. Sindermann
Major Marine Ecological Disturbances, Ernest H. Williams, Jr. and Lucy Bunkley-Williams
Marine Bivalves and Ecosystem Processes, Richard F. Dame
Morphodynamics of the Inner Continental Shelf, L. Donelson Wright
Practical Handbook of Marine Science, 2nd Edition, Michael J. Kennish
Seabed Instability, M. Shamim Rahman
Sediment Studies of River Mouths, Tidal Flats, and Coastal Lagoons, Doeke Eisma

---

PREFACE

For more than 40 years, the *Calypso* (Cousteau) team has been exploring the coral reefs of the world — including those of the Caribbean and the Red Sea as well as the very rich Indonesian reefs and, of course, the Great Barrier Reef in Australia. After a time of marveling, the Cousteau team very soon realized that alterations caused by the anarchic development of the human species were so quick that coral reefs, developing at the pace of geological eras, could not survive in the long run.

In rich countries, the leisure industry keeps selling "coral dreams". Unscrupulous persons remodel the coastline according to their whims. They build marinas or airports right where reefs lie, thus destroying the very beauty that they wanted to enjoy. In poor countries, overpopulation and malnutrition force people to devastate their land and sea resources in the interest of day-to-day survival. We have witnessed such tragedies in Haiti and also in Malaysia, where children, at the risk of their lives, use dynamite to blow up reefs that once fed their parents and that should still be their gardens.

The Great Barrier Reef, better protected by a legal framework which the Australian government was wise enough to impose, seemed wonderful to the Cousteau team when we arrived to watch the mass reproduction of coral. But here and there we were aware of covert threats posed to this jewel through the action of property developers and hordes of tourists.

In June 1992, at the U.N. Conference on Environment and Development held in Rio de Janeiro, delegates representing all the nations of the world signed a convention through which signatories committed themselves to preserve biodiversity so as to fulfill the needs of present and future generations. Protection of any species cannot be envisioned outside of the ecosystems they constitute. With 350 species of coral and 1500 species of fish, the Great Barrier Reef is one of the richest ecosystems in the world. It is high time that we acquire the scientific means to preserve it.

This reference work by Dr. Eric Wolanski stands as the essential synthesis with respect to physical mechanisms conditioning life in this very complex ecosystem. It will doubtlessly help the Great Barrier Reef Marine Park Authority ensure the preservation of the eighth wonder of the world for future generations.

Francois Sarano
Equipe Cousteau

THE AUTHOR

Eric Wolanski, Ph.D., is a Senior Principal Research Scientist at the Australian Institute of Marine Science, Townsville, Australia.

Dr. Wolanski graduated in 1969 from the Catholic University of Louvain, Louvain, Belgium, with a B.Sc. degree in civil engineering. He obtained his M.Sc. degree in civil and geological engineering in 1970 from Princeton University, Princeton, New Jersey, and his Ph.D. degree in environmental engineering in 1972 from the Johns Hopkins University, Baltimore, Maryland. He was a postdoctoral fellow at the California Institute of Technology. He worked in Australia at the Bureau of Meteorology, the State Pollution Control Commission, and the Snowy Mountains Engineering Corp. before joining the Australian Institute of Marine Science in 1978.

He is a member of the Sigma Xi Research Society of North America and the Australian Institution of Engineers, and a foreign correspondent of the Belgian Royal Academy of Overseas Sciences. He is a member of the editorial advisory boards of the scientific journals *Estuarine, Coastal and Shelf Science, Continental Shelf Research, Journal of Coastal Research, Journal of Marine Systems,* and *Oceanographic Literature Review*. He has published more than 120 research papers. His current major research interests include the water circulation in a topographically complex environment and its implications for dispersion and mixing, siltation processes, and interactions between physical and biological processes.

TABLE OF CONTENTS

Introduction

My interest in coral reefs and the circulation around them was first aroused in 1961 when I had occasion to visit Tahiti and neighboring islands and to swim and dive around them. I observed a few features of the circulation in the lagoons but on returning home and turning to the literature I was surprised to find out how few physical oceanographic studies had been made in such waters. In subsequent years I have visited some 60 islands in tropical waters in the three oceans and so enlarged my experience of reef areas, visiting the Great Barrier Reef first in 1971. In 1975 I accepted an invitation from the Australian Institute of Marine Science (A.I.M.S.) to prepare an account of such physical oceanographic information as was then available about the Great Barrier Reef and western Coral Sea as background for the biologists and geologists who formed the staff at the time. The information available was limited to a modest climatology of the region, some information on tides, a little on currents near the major ports, and some on the water properties, all of it descriptive in character and distinctly limited in extent in both space and time. Since that time I have been fortunate in having the opportunity to visit A.I.M.S. a number of times to collaborate in ongoing research and even to carry out one study of my own (in support of the MECOR project).

In 1978, Dr. Eric Wolanski joined A.I.M.S. as its first physical oceanographer. His early work there included some more detailed synoptic observations of water properties in the area near Townsville but he soon started to direct his attention to the more urgent need to learn more about the circulation and its dynamics, i.e., the forces acting on the water and the resultant motions in the complicated topography of the continental shelf infested with coral reefs of a very wide range of sizes and shapes.

An early and most important result in 1983 was the discovery of the occurrence of continental shelf waves of periods of several days and the recognition of the driving force, i.e., the periodic variations of the southeast trade winds. The identification of these waves was important because they have only a small vertical amplitude and generate modest currents, so that they required intensive current metre studies to detect them, but their long periods of 5 to 10 d mean that horizontal excursions of the waters and their contents, such as plankton, eggs, pollutants, etc., can be very significant, of the order of 100 to 300 km. Wolanski and colleagues, and Middleton and his, made extensive field and theoretical studies of these long waves. They are a common feature of continental shelf waters around the world, but the presence of the extensive reefs complicates their study in the Great Barrier Reef area.

Another consequence of the presence of reefs on the shelf is the occurrence of wakes, eddies, and turbulence which can occur in the tidal streams behind such obstructions. These features have been studied in a number of locations in the northern and central regions, with Wolanski's densely instrumented study around Rattray Island revealing significant features of the horizontal and vertical motions in such wakes and providing field data for the testing of a number of mathematical models of flows near reef obstacles. The effects of reefs in focusing waves, previously qualitatively recognized by geologists, is now being studied using such mathematical models.

As in most other aspects of meteorology and oceanography, mathematical models play a very significant role in the study of reef water circulations, both with the aim of understanding the relative significance of driving and friction forces and also for the prediction of features such as the spread of pollutants, distributions of planktonic eggs and spawn, and fishes. A number of models have been designed to predict local area and more extensive area circulations in normal circumstances and also in the less regular occurrence of strong forcing during cyclones.

The occurrence of gaps between pairs of reefs, particularly between the long Ribbon Reefs at the eastern edge of the reef in the north, gives rise to jets during tidal flood and ebb. During the flood these jets can cause upwelling, in the Coral Sea nearby, of relatively nutrient-rich subsurface water which is carried into the shallower reef area as a major source of nutrients, while ebb current jets can modify the water property distributions in the Coral Sea outside the reefs (and on one occasion hindered Cook in returning to the Reef Lagoon from the Coral Sea).

Measurements along the outer edge of the reef near the shelf/slope boundary have shown that large vertical oscillations of tidal period ("internal tides") can take place, which also bring the deeper water onto the shelf and possibly into the reef area. Analyses of circulation measurements in a number of areas has well demonstrated the effect of the East Australian Current in causing a general southward flow in the central and southern regions of the reef, although this flow may be overridden at times by the wind-induced circulations.

A feature of the distributions of water properties which has been revealed by both *in situ* observations and satellite observations is that patchiness of properties is quite common. The distribution of fresh water from river runoff into the reef area is made conspicuous, by its sediment content, to visual observation, which has shown its expected tendency to remain close to the shore in its northward progress but the reason why such flows eventually break up into "clouds" is not known. Other dynamic processes have been demonstrated to give rise to patchy distributions of water properties.

An unusual contribution by Wolanski and associates has been the application of physical oceanographic methods to study the circulation in mangrove swamps which occupy a significant proportion of the shore sided of the reef. These regions are well known to be a significant source of nutrients to the marine waters outside, and therefore to the biota there, but few studies had been made previously of the complex circulations in such regions.

The above remarks outline some of the applications of physical oceanographic studies, synoptic and dynamic, which have been made in the Great Barrier Reef region and which have revealed a wide range of physical features of the circulation previously unsuspected. All this has been accomplished in the brief period of less than 15 years since my summary in 1976 found only simple descriptive material available. Eric Wolanski is to be congratulated for summarizing in his account, the wide range of studies by many Australian and visiting oceanographers which have made the Great Barrier Reef of Australia by far the best described physically of any reef area in the world. The extent of his own and his colleagues' contributions is evident in the list of references to published work.

George L. Pickard
Former Head
Department of Oceanography
University of British Columbia

1 Geographic Setting

The Great Barrier Reef (Figure 1.1) extends approximately 2600 km along the eastern coast of Australia, from just north of Fraser Island in the south (25°S) to the coast of Papua New Guinea in the north (9.2°S). It is not a continuous barrier. Instead, it is a matrix of 2500 individual reefs (Figure 1.2). The reefs range in size from 100 km^2 to one-hundredth of 1 km^2. The reefs are surrounded by interreefal waters forming channels. The channels can be only several hundreds of metres wide; in other cases they are tens of kilometres wide. The assemblage of reefs and interreefal waters is called the Great Barrier Reef matrix. This matrix is located on a continental shelf. For convenience, the Great Barrier Reef is divided into four regions.

Torres Strait (Figure 1.2d) is located north of Cape York and is influenced by the Gulf of Papua, the Gulf of Carpentaria, and the Coral Sea. The shelf is 100 km wide, shallow (depth seldom exceeding 20 m), and largely uncharted. The western side of Torres Strait faces the Gulf of Carpentaria. In this region, the water depth is seldom greater than 15 m. The northern half is no more than 5 to 10 m deep, with numerous shoals and reefs shown as stipled areas in Figure 1.2d. There are narrow channels with violent currents (Figure 1.3). The Torres Strait is also open to the Gulf of Papua through the Great North East Channel. The channel is marked on the west by the Warrior Reefs and to the east by a dense matrix of reefs. The bottom of the Gulf of Papua slopes gently to a shelf break located about 100 km from the coast.

The Northern Region (Figure 1.2c) includes all the reefs to the north of 16°S. The waters are very shallow, with a maximum depth seldom exceeding 30 m. A number of reefs and shoals are scattered very densely all across the shelf width (Figure 1.4). Elongate reefs are found at the shelf break, separated by narrow reef passages. This area is generally poorly charted except in the navigation channels.

The Central Region (Figure 1.2b) extends from about 16°S to about 20°S. The sea floor is fairly flat in the northern sector of this region, with the maximum depth rarely exceeding 40 m. In the central sector off Townsville (19.2°S), the sea floor slopes gently from the coast to about 100 m at the shelf break. This area, which includes Palm Passage, has the least reef density of the whole Great Barrier Reef. There are several large rivers in this area (Figure 1.5). Most reefs are located

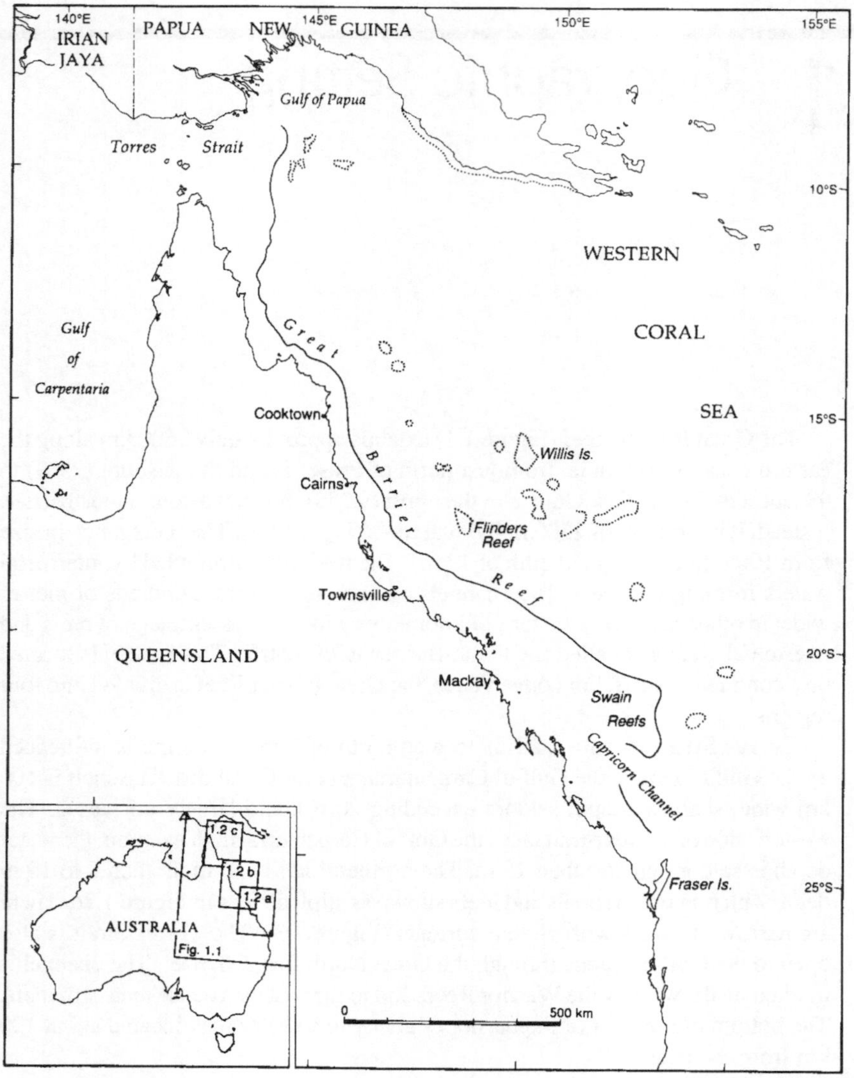

FIGURE 1.1. General location map.

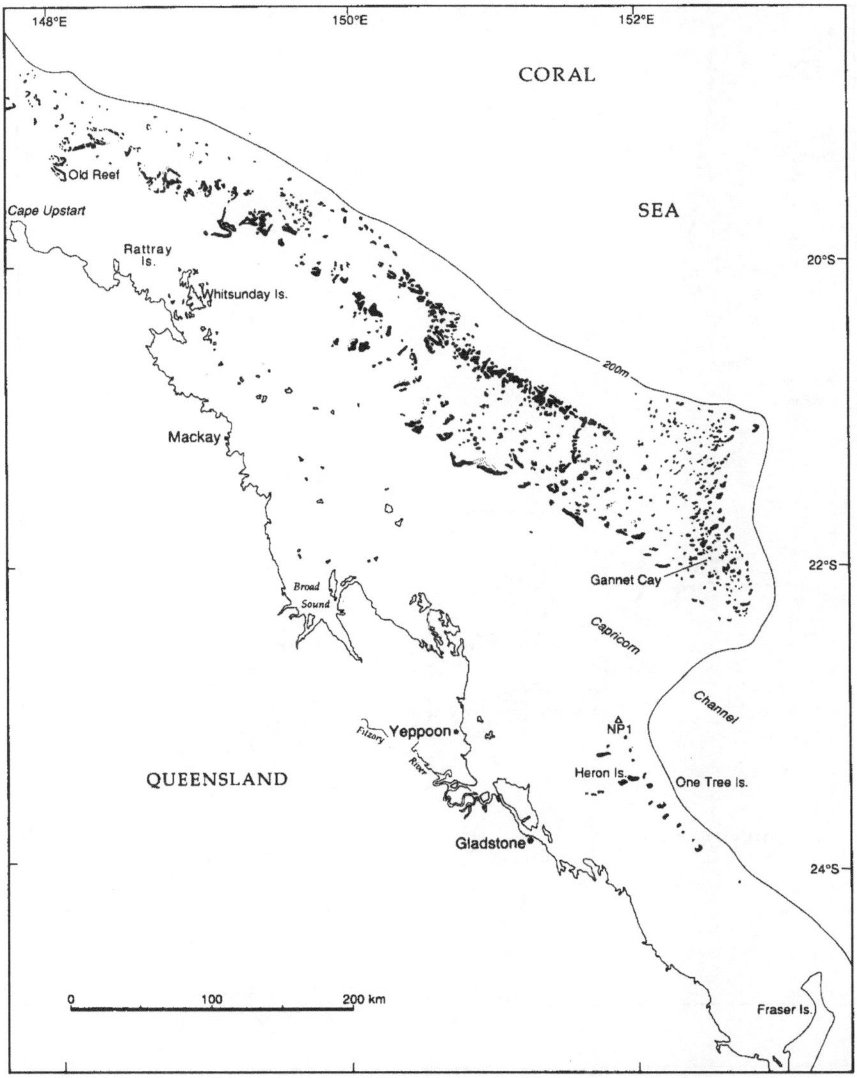

FIGURE 1.2. Location maps of the (a) northern, (b) central, and (c) southern regions, and (d) Torres Strait with mooring sites. (Map d is adapted from Wolanski et al., 1988b.)

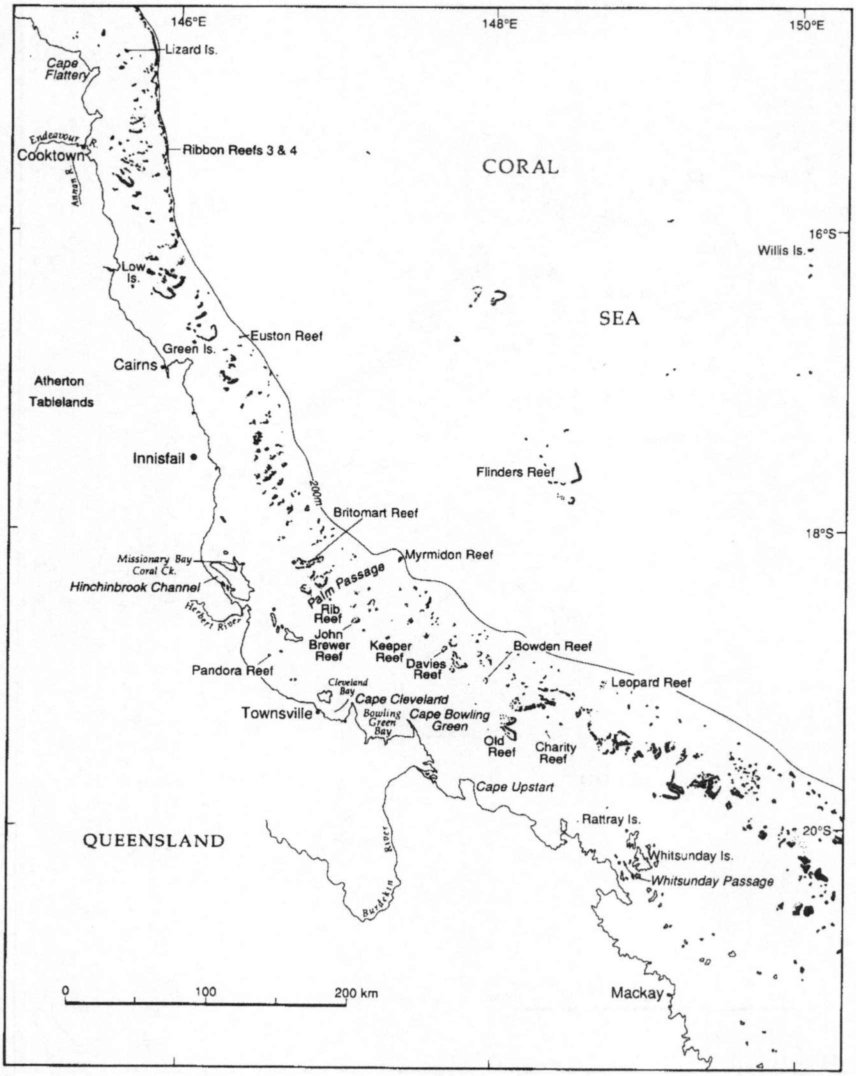
146°E
148°E
150°E
16°S
18°S
20°S
Lizard Is.
Cape Flattery
Endeavour R.
Cooktown
Annan R.
Ribbon Reefs 3 & 4
CORAL
SEA
Willis Is.
Low Is.
Euston Reef
Green Is.
Cairns
Atherton Tablelands
Innisfail
Flinders Reef
200m
Britomart Reef
Myrmidon Reef
Missionary Bay
Coral Ck.
Hinchinbrook Channel
Herbert River
Palm Passage
Rib Reef
John Brewer Reef
Keeper Reef
Davies Reef
Bowden Reef
Pandora Reef
Cleveland Bay
Cape Cleveland
Leopard Reef
Townsville
Bowling Green Bay
Cape Bowling Green
Old Reef
Charity Reef
Cape Upstart
Rattray Is.
QUEENSLAND
Burdekin River
Whitsunday Is.
Whitsunday Passage
Mackay
0
100
200 km

FIGURE 1.2b.

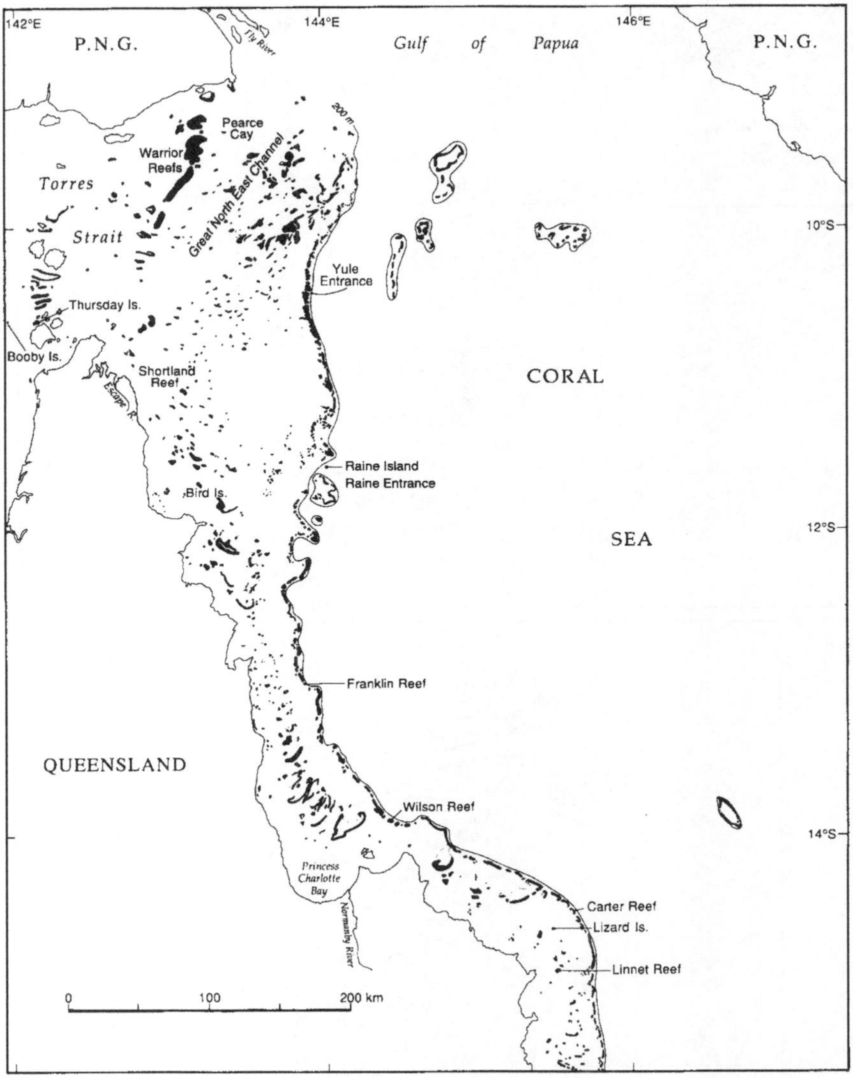
142°E
144°E
146°E
P.N.G.
Gulf of Papua
P.N.G.
Fly River
200 m
Pearce Cay
Warrior Reefs
Great North East Channel
Torres
Strait
10°S
Yule Entrance
Thursday Is.
Booby Is.
Shortland Reef
Escape R.
CORAL
Raine Island
Raine Entrance
Bird Is.
12°S
SEA
Franklin Reef
QUEENSLAND
Wilson Reef
14°S
Princess Charlotte Bay
Normanby River
Carter Reef
Lizard Is.
Linnet Reef
0
100
200 km

FIGURE 1.2c.

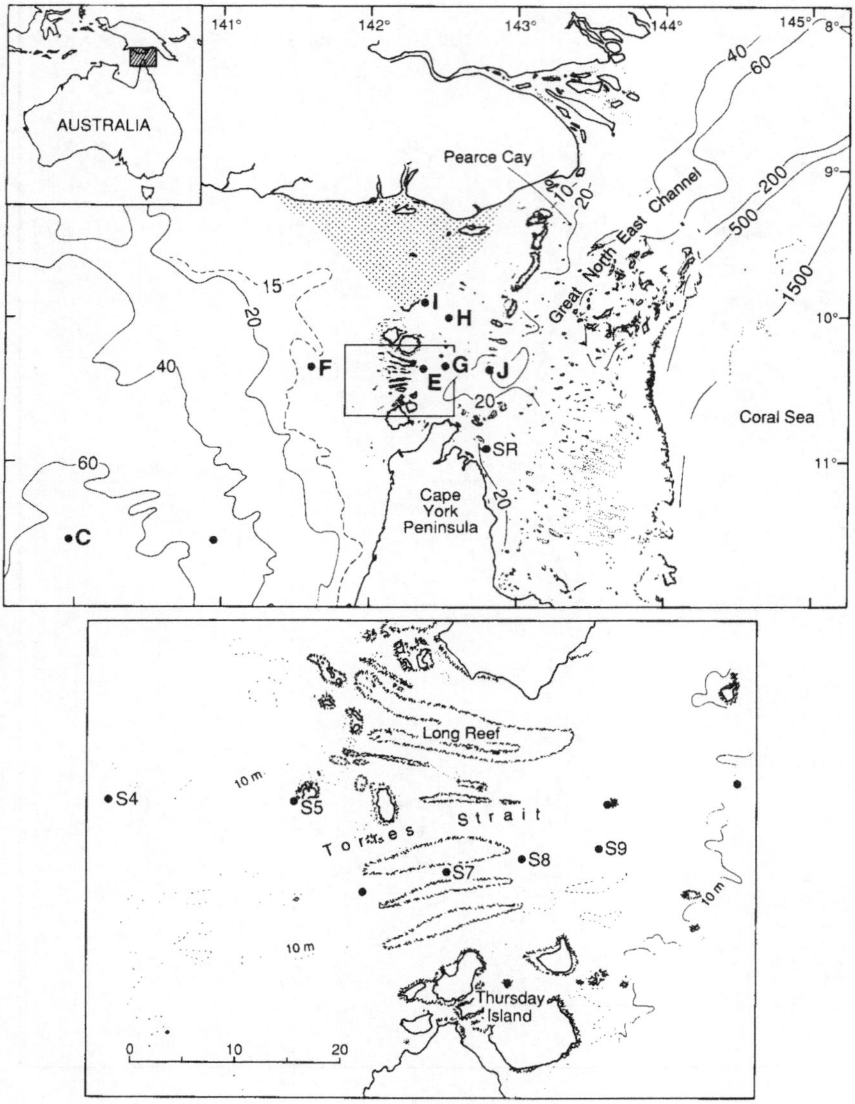
141°
142°
143°
144°
145°
8°
9°
10°
11°
AUSTRALIA
Pearce Cay
Great North East Channel
Coral Sea
Cape York Peninsula
10
15
20
40
60
200
500
1500
•C
•F
•I
•H
•G
•E
•J
•SR
Long Reef
Torres Strait
Thursday Island
10 m
•S4
•S5
•S7
•S8
•S9
0
10
20

FIGURE 1.2d.

FIGURE 1.3. Strong currents through Torres Strait.

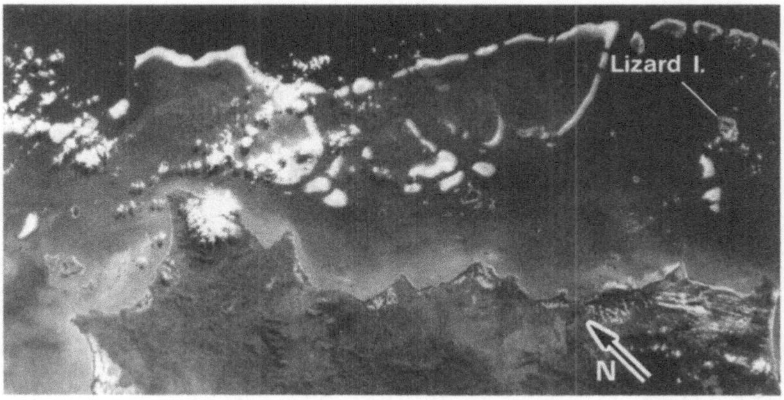

FIGURE 1.4. Raw Landsat band 4 view of the continental shelf at 14.5°S.

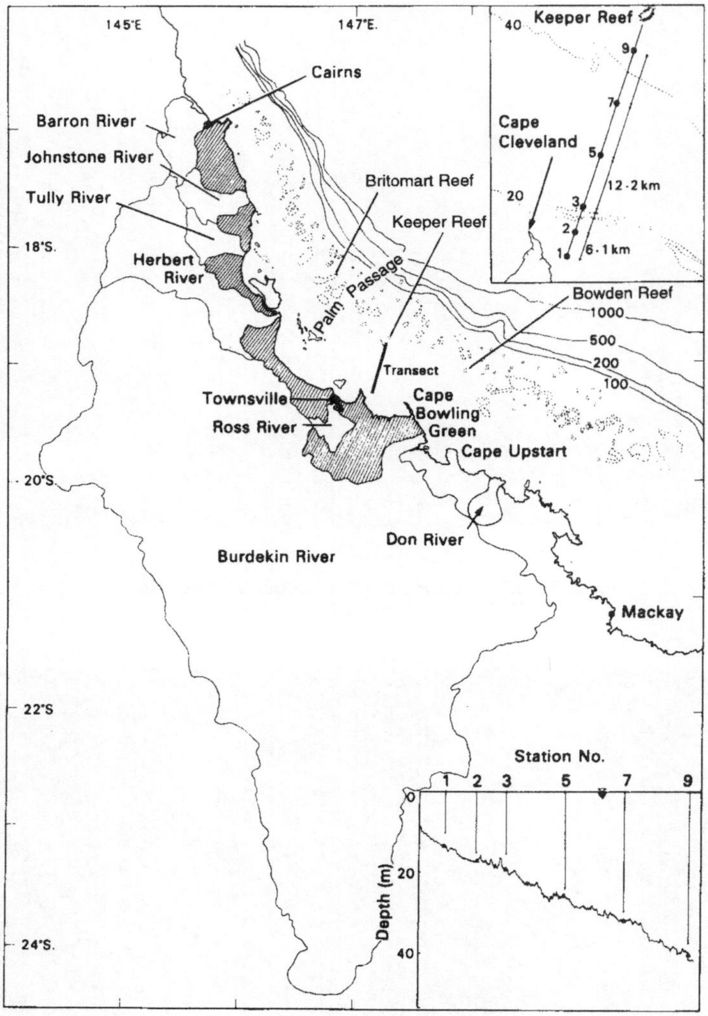

FIGURE 1.5. Map of the central region (depths in metres) and (inset) bathymetric details of the Keeper Reef transect. (Adapted from Wolanski et al., 1981.)

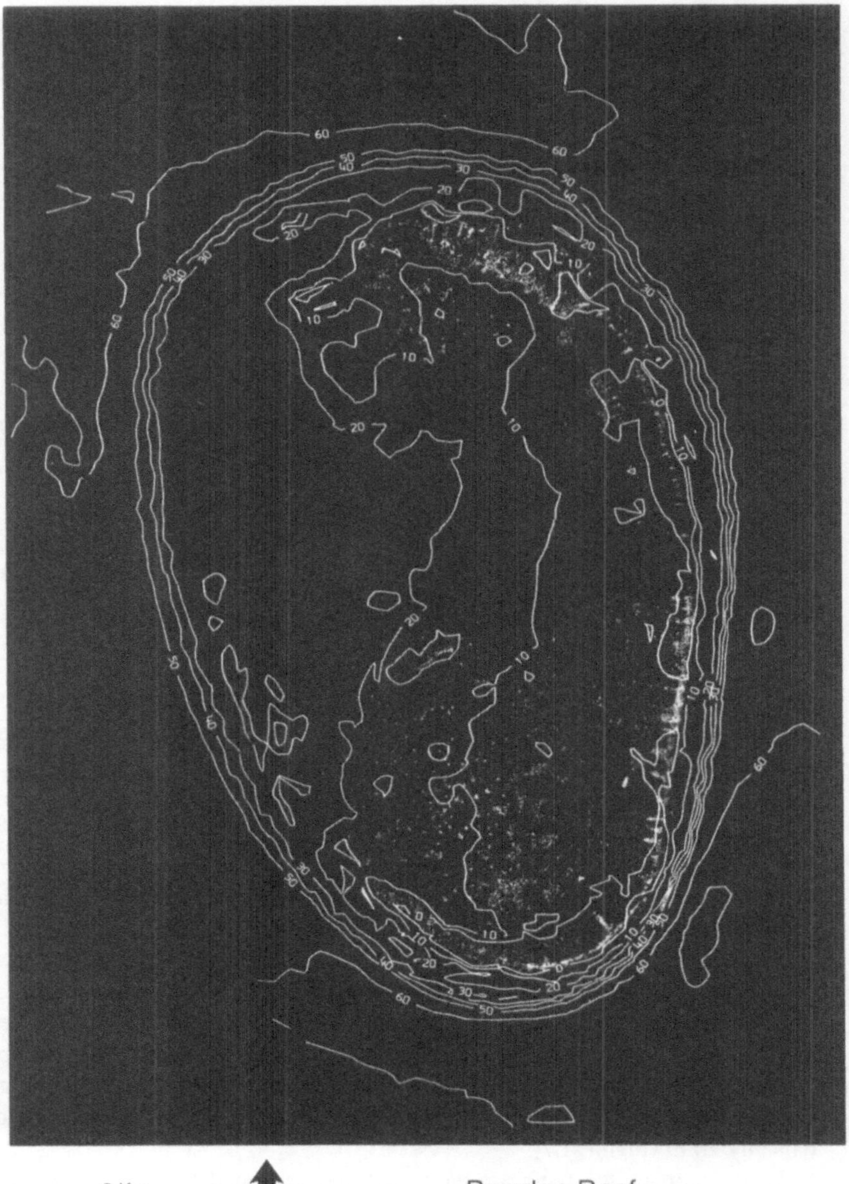

FIGURE 1.6. Bathymetry (in metres) of Bowden Reef.

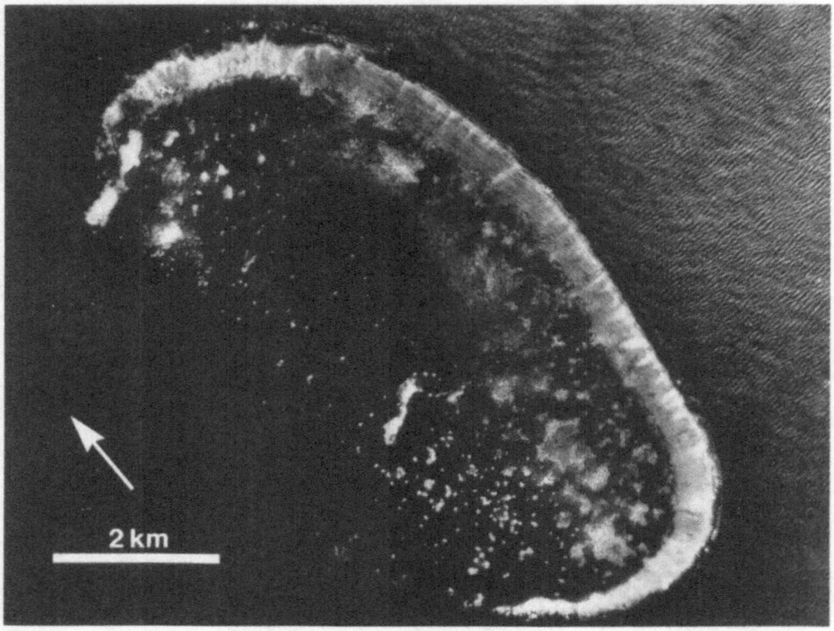

FIGURE 1.7. Aerial photograph of Bowden Reef.

on the mid- and outer shelf and only the inner shelf is reef free. The coast is very rugged, with numerous headlands sheltering shallow (a few metres deep) mangrove-fringed embayments.

The Southern Region (Figure 1.2a) extends south of 20°S. Water depth reaches 145 m. In the northern sector, the reef matrix is very dense. In the southern region, there is the Swain complex of small patch reefs. This area is open to the Coral Sea through the Capricorn Channel. Both the mid-shelf and the inner shelf are largely reef-free.

The reefs in all regions vary in shape from kidney-shaped reefs with a lagoon, to the elongate Ribbon Reefs, to flat platform reefs without a lagoon. Bowden Reef (19°S) is a typical kidney-shaped reef (Figure 1.6). The windward reef slope is usually very steep. At the lee side of the reef and in the lagoon, there are numerous small coral outcrops reaching all the way to the surface from depths of typically 20 m (Figure 1.7).

In some areas, the reefs are elongated and separated by narrow passages typically 40 m deep (Figure 1.8). On the shelf extensive meadows of the alga *Halimeda* are found, forming banks rising 10 to 20 m above the surrounding sea floor. There are also mid-shelf reefs of various shapes.

The surface of the reef is very rugged, with a rugosity that can vary from a few centimetres on a heavily cemeted reef flat to several metres in areas of prolific reef growth.

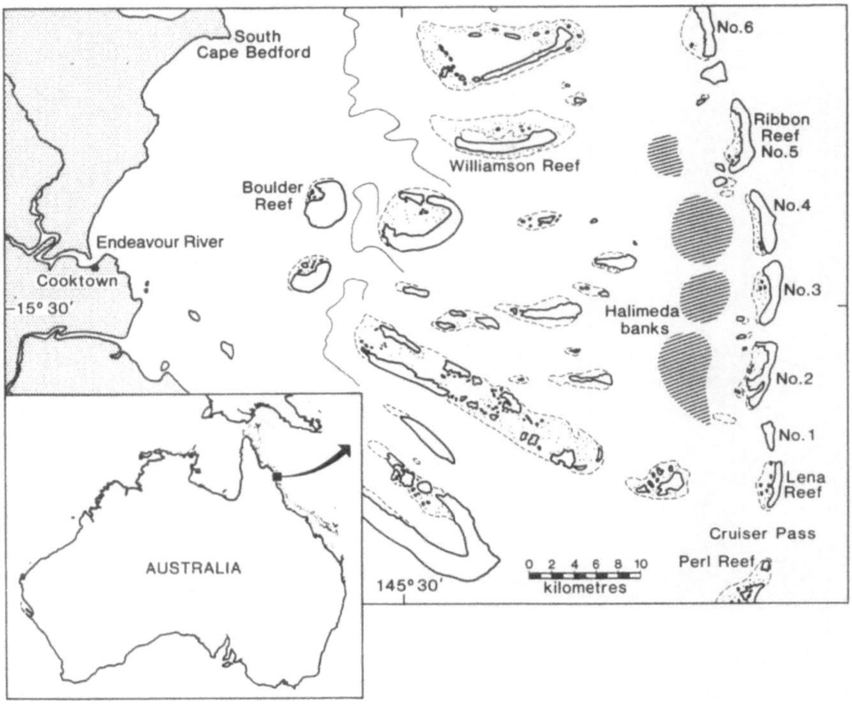

FIGURE 1.8. Map of the region around 15.5°S showing the Ribbon Reefs at the edge of the continental shelf, meadows of the alga *Halimeda* on the outer shelf, and mid-shelf reefs. (Adapted from Wolanski et al., 1988a.)

2 Wind, Rainfall, and Sediment Inflow

2.1 METEOROLOGY

The Bureau of Meteorology and the A.I.M.S. maintain an impressive network of weather stations along the coast and on islands and coral cays on the continental shelf and in the Coral Sea (Figure 2.1). Most of these stations transmit weather data by telemetry and in near-real time. In addition, upper-air observations are also taken at some stations, as well as weather observations by radar.

Rainfall and river heights are also measured by the Bureau of Meteorology and the Queensland Water Resources Commission; together they operate about 580 river-height stations and 700 rainfall stations. However, most of these rainfall stations are located on the mainland because problems with fouling by birds (guano etc.) preclude the reliable use of automatic telemetric rainfall stations in the reef area.

The dominant wind is the southeast trade wind which prevails from about April to October but can occur also at other times. Southeast trade winds seldom exceed 30 knots (15 m s^{-1}). In summer, the winds are more variable in direction and generally weaker except in tropical cyclones. Figure 2.2 shows wind roses at selected coastal and offshore stations. In July (winter), the southeast trade winds dominate. In January (summer), the wind is more variable. The weather is then influenced by the shift into the Southern Hemisphere of the Intertropical Convergence Zone and the southwestward shift in summer of the south Pacific Convergence Zone. A northwest wind then prevails in the northern Great Barrier Reef, with speed seldom exceeding 15 knots (7.5 m s^{-1}). Trade winds can still occur in summer but they then have a more easterly component.

In winter the trade winds are generally coherent over the whole Great Barrier Reef (Figure 2.3). Figure 2.3 shows that the wind fluctuates with periods of several days to several weeks. These fluctuations are very coherent throughout the region in the southeast trade wind season when they are generated by the passage of weather systems over the southern part of Australia. The time lag between the passage of the weather system at these various stations is small, e.g., winds at Thursday Island lag those at Rib Reef by only 13 h. Autospectra (Figure 2.4, line

FIGURE 2.1. Meteorological stations.

a) of the longshore wind component are energetic at all periods >5 d, with a number of peaks centered at about 9 to 12 d (the "weather band"), 50 to 70 d, and 180 d (the "seasonal band").

The sea breeze is pronounced at coastal and some offshore stations, but generally not at stations located more than 80 km from the coast. The seabreeze commonly results at coastal stations in a 45° anticlockwise change in wind direction between 0900 and 1500 h.

Tropical cyclones (hurricanes, typhoons) can occur anywhere, though they are rare north of Thursday Island (10.5°S). The tropical cyclone season lasts from November to May, but 75% of the cyclones occur in January and February (Downey, 1983). About two or three tropical cyclones a year make landfall. The

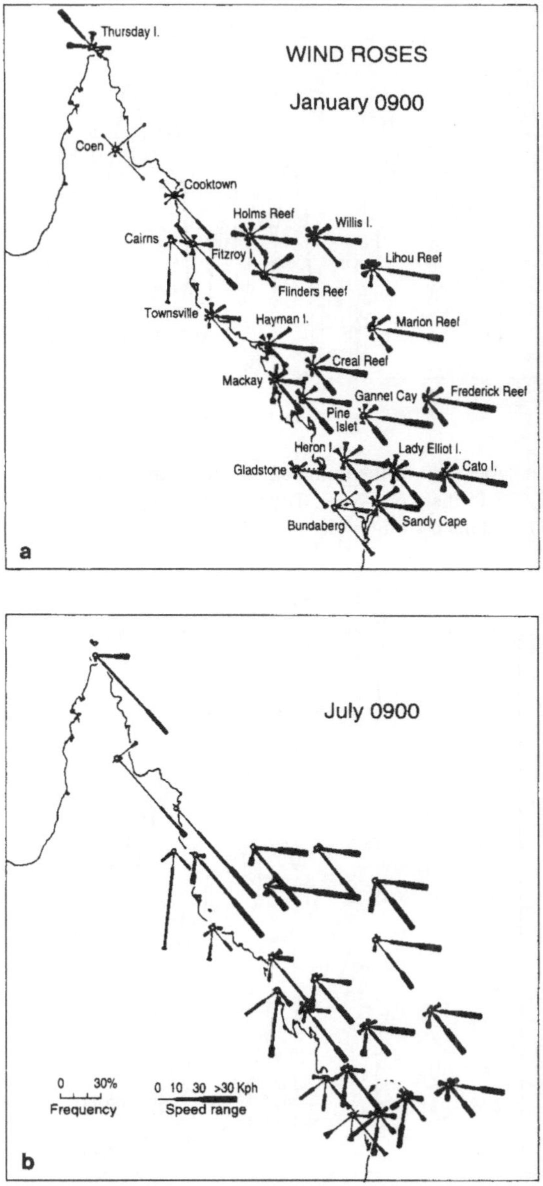

FIGURE 2.2. Wind roses at 0900 h in (a) January and (b) July. (Adapted from Downey, 1983.)

return period for landfall is about 96 years for any 100-km section in the central region, but this varies with latitude. The probability of landfall shows a broad maximum between 15 and 17°S. Several vortex models are available to calculate the wind field generated by a tropical cyclone. These models calculate the wind

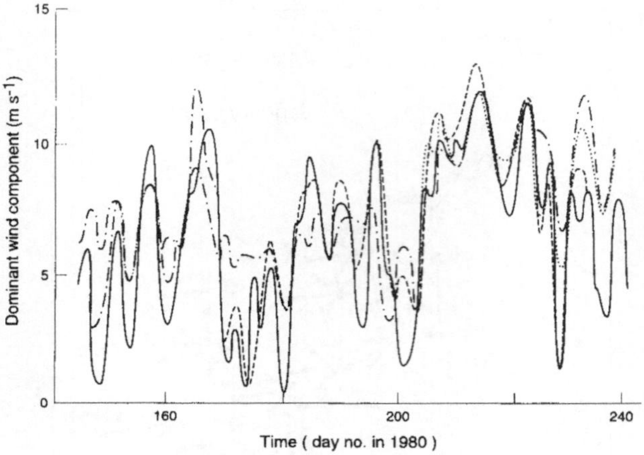

FIGURE 2.3. Time series plot of the dominant wind component from June to October 1980 at Flinders Reef (17.7°S in the Coral Sea), Thursday Island (10.5°S), Rib Reef (18.5°S), and Carter Reef (14.6°S). (Adapted from Wolanski, 1982.)

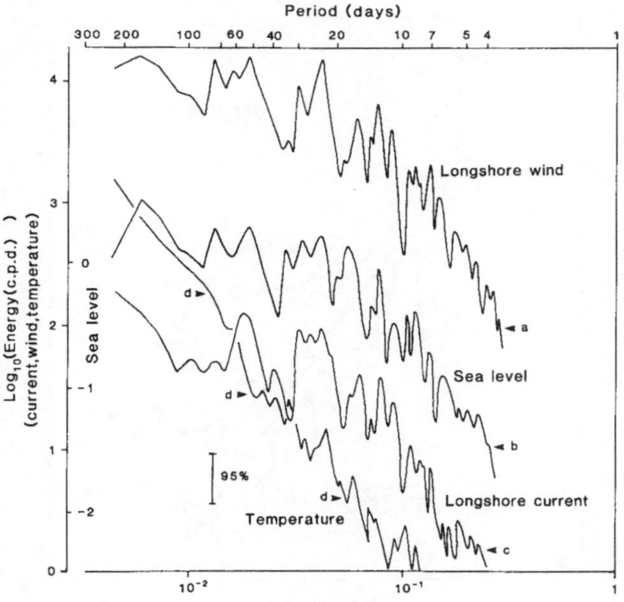

FIGURE 2.4. Autospectra of (a) the longshore wind component at Rib Reef, (b) the low-frequency sea level at Townsville, (c) the low-frequency longshore current off Green Island, and (d) the temperature off Green Island. (From Wolanski and Pickard, 1985. Reproduced with the kind permission of Springer-Verlag.)

field from the central pressure (p_c, in hPa), radius of maximum winds (R, in metres), and the forward velocity of the cyclone (V_f, in metres per second). One of the most sophisticated is that of Graham and Hudson (1960). A semi-empirical formula exists to estimate the maximum 10-min mean wind, U_r (in metres per second) (Kjerfve and Dinnel, 1983; Kjerfve et al., 1986; Walker et al., 1988)

$$U_r = C(p_a - p_c)^5\left(\frac{R}{r}\right)^k + 1.17\ V_f^{0.63}\cos\theta/U_{rs} \qquad (2.1)$$

where U_{rs} is the maximum 10-min mean wind speed in a stationary cyclone, r is the distance from the eye of the tropical cyclone, p_a an ambient pressure (that can be set a 1010 hPa), θ the angle in a clockwise direction from the eye to the site location, and C and k are constants with typical values of 4.3 and 0.67, respectively. Done (1992) calculated from Equation 2.1 the wind at Lizard Island (14.6°S) under tropical cyclone 'Ivor' on 19–20 March 1990 from radar estimates of the cyclone track (Figure 2.5a) and p_c. The comparison between observed and predicted 10-min mean winds (Figure 2.5b) is quite encouraging for the wind speed, less so for the wind direction.

2.2 FRESHWATER INPUT

Freshwater enters the Great Barrier Reef via direct rainfall and from river discharges. No data are available on underground inflow, if any.

2.2.1 Rainfall

Monthly rainfall on islands varies seasonally and with latitude (Figure 2.6a). Monthly rainfall is minimal in winter and maximal in the summer. About 60 to 70% of the rainfall occurs from January to March. There is also considerable interannual variability (Figure 2.6b). Typically the standard deviation of the annual rainfall is 60% of the mean. This phenomenon is linked to both the erratic nature of tropical cyclones and the El Niño phenomenon.

2.2.2 Freshwater Inflow

The three largest rivers in the Great Barrier Reef are the Burdekin River in the central region (Figure 1.5), the Fitzroy River in the south, and the Normanby River in the north. The Burdekin River is the largest, in terms of mean annual flow (9.7×10^9 m^3) and in terms of land drainage area (1.3×10^5 km^2). There are also a number of smaller coastal streams.

The river discharges vary considerably, with time scales of hours (for small coastal streams), days (for the largest rivers, such as the Burdekin River), months

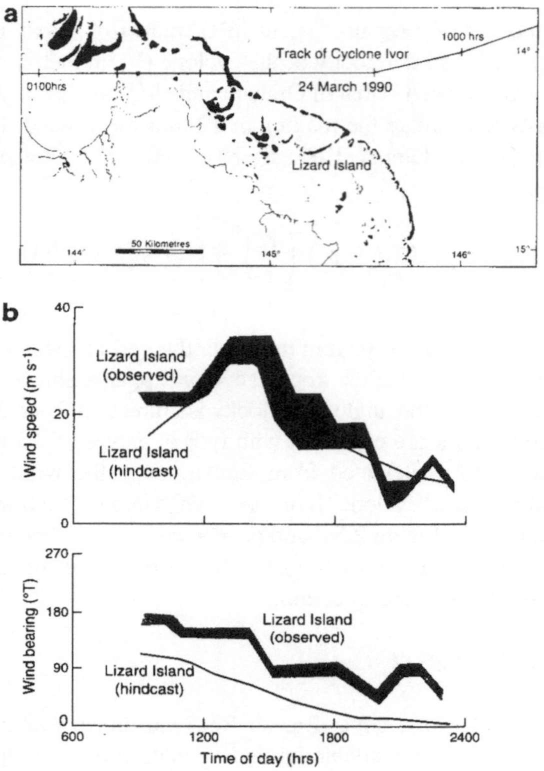

FIGURE 2.5. (a) Location map and cyclone track and (b) observed and hindcast wind speed and direction at Lizard Island. (Adapted from Done, 1992.)

(seasonal variations), and years (interannual variability, strongly linked to tropical cyclones and the El Niño phenomenon). Thus, most of the freshwater discharge occurs in a few short-lived floods in summer. For several months in winter, most river beds are commonly dry, with occasional pools of impounded water in scour holes. For small streams such as the Annan River (15.5°S) with a drainage area of 300 km^2 above the gauging site, a flood lasts only a few hours (Figure 2.7a).

The discharge of the Burdekin River also fluctuates in a similar way although the flood rise and fall last longer, typically a few days. As an example, Figure 2.7b shows a time series of the discharge of the Burdekin River in the wet season of 1981. There were three distinct flood events, including two dominant ones. Some years no floods occur, most years only one flood event. Burdekin River peak discharges are typically on the order of 10,000 to 15,000 m^3 s^{-1}. The maximum possible discharge exceeds 30,000 m^3 s^{-1}. The bulk of the discharge of the Burdekin River (Figure 2.8) occurs in the wet season from January to March, a pattern clearly reflecting the seasonal variation of the rainfall. There is also considerable interannual variability. The analysis of ring patterns in cores from coastal corals

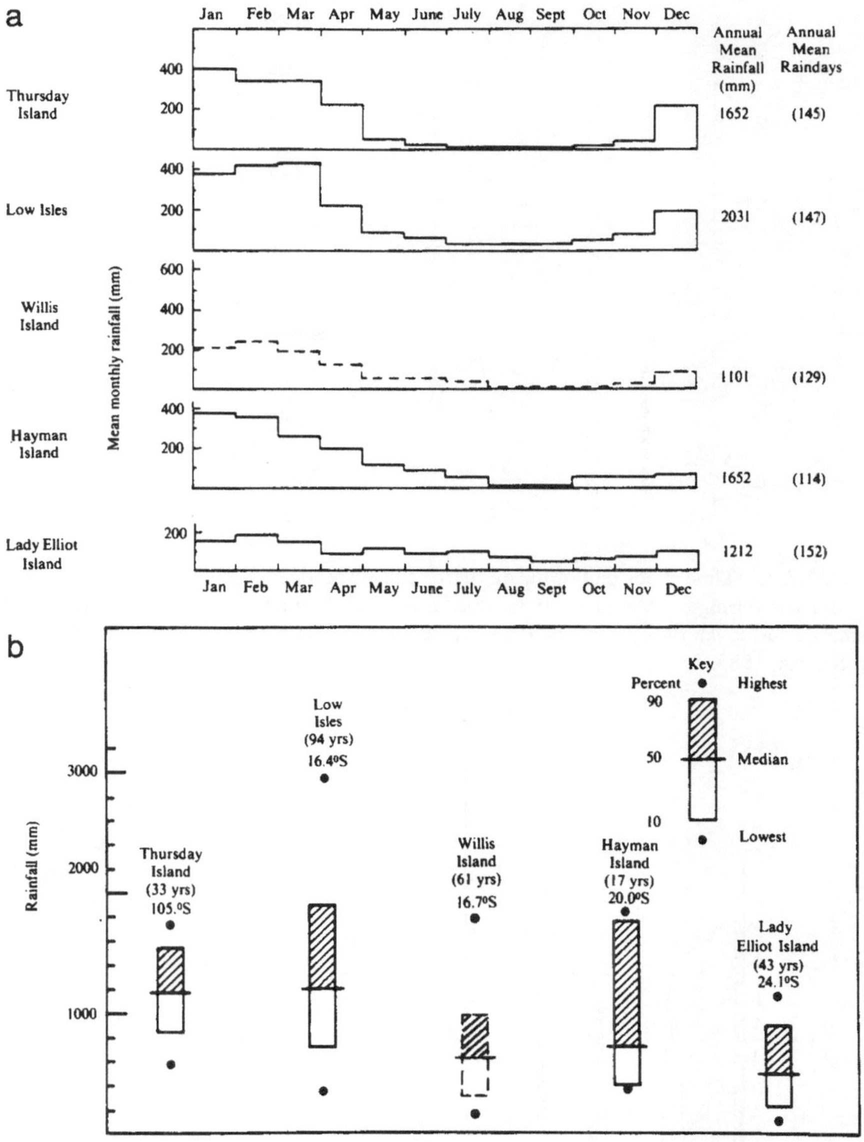

FIGURE 2.6. (a) Monthly mean rainfall at various stations and (b) standard deviation. (Adapted from Downey, 1983.)

(Isdale, 1984; Boto and Isdale, 1985) has enabled the calculation of the approximate monthly discharge of the Burdekin River for over 200 years. These predictions suggest a variable cycle of years of alternately drier and wetter weather.

In the Gulf of Papua, the Fly River discharge is about 6,000 $m^3 s^{-1}$, with little seasonal fluctuations.

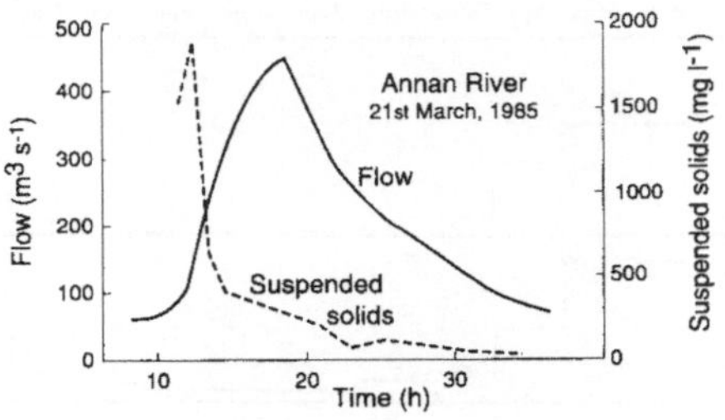

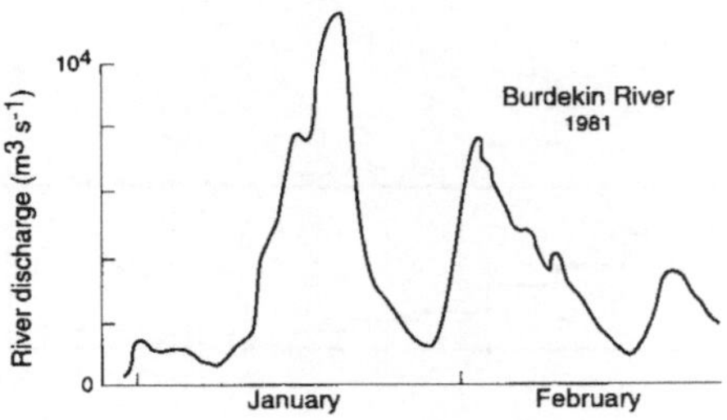

FIGURE 2.7. Time series plot of the discharge and suspended sediment concentration in the Annan River during the March 21, 1985, flood event and (b) time series of the Burdekin River discharge during the flood events of 1981. (Adapted from Hart et al., 1988, and Wolanski and van Senden, 1983.)

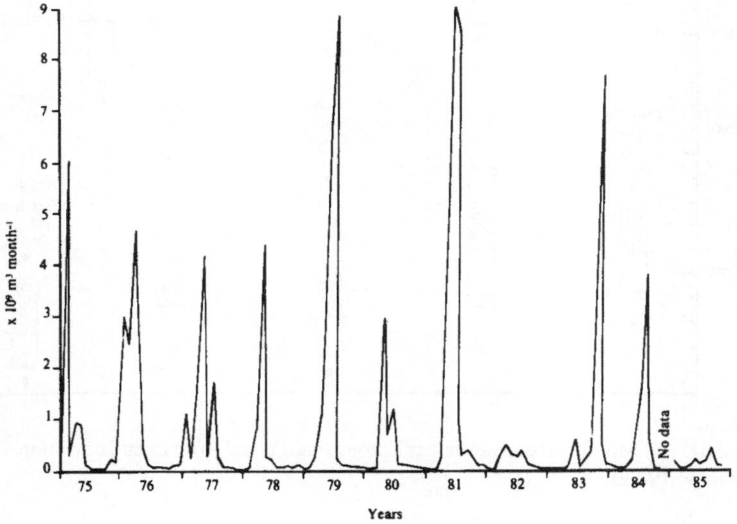

FIGURE 2.8. Monthly run-off of the Burdekin River from 1975 to 1985.

2.2.3 Estuarine Filtering

The river inflow at the head of an estuary may be significantly modified during the advection through the estuary (Kennedy, 1984). This filter effect is particularly strong in mangrove-fringed estuaries, which are common along the Great Barrier Reef.

In the dry season, evapotranspiration from mangroves withdraws freshwater from saline water at a rate of typically 0.8 cm d^{-1} and creates a salinity maximum zone in the estuaries that can effectively isolate the river from the coastal waters (Figure 2.9).

Lateral trapping is also enhanced by mangroves. Lateral trapping was first studied by Okubo (1973) for an estuary with a lateral embayment (Figure 2.10). When a tagged water mass moves upstream at flood tide, a fraction of that water ends up trapped in the lateral embayment. At ebb tide, this mass returns to the tidal creek and mixes with untagged water. Mixing is thus greatly enhanced. This effect can be parameterized by a tide-averaged longitudinal eddy diffusivity

$$B = \frac{A}{1+\varepsilon} + \varepsilon \; U_o^2 / 2k\,(1+\varepsilon)^2 (1+\varepsilon+\sigma/k) \tag{2.2}$$

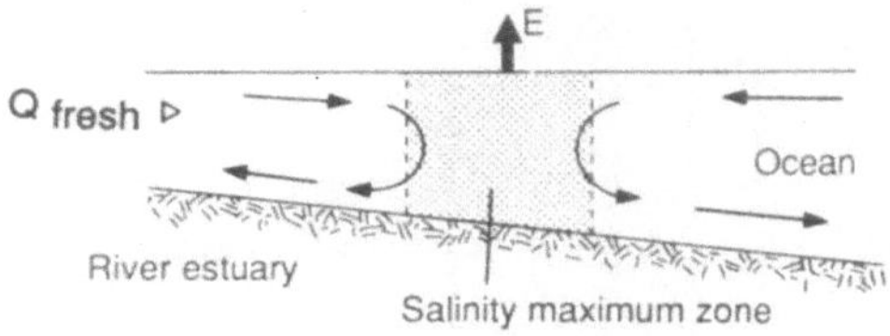

FIGURE 2.9. Internal circulation in the dry season in a mangrove-fringed estuary. (Adapted from Wolanski, 1986a.)

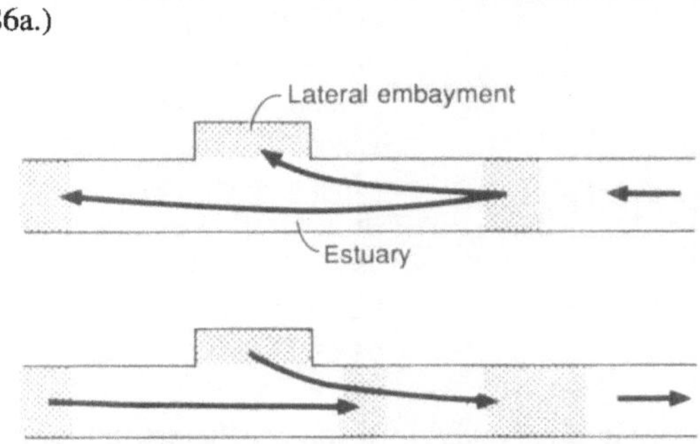

FIGURE 2.10. Sketch of the lateral trapping phenomenon in a tidal estuary with an embayment. (Adapted from Okubo, 1973.)

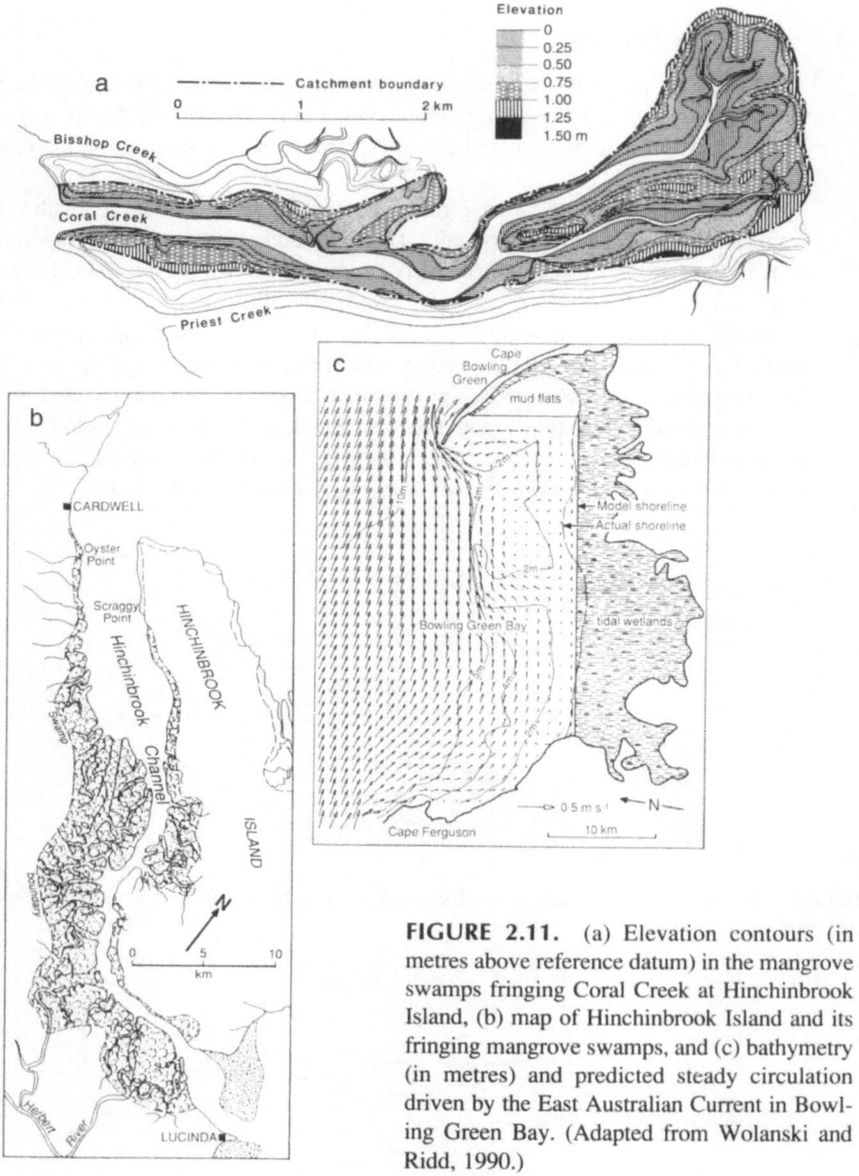

FIGURE 2.11. (a) Elevation contours (in metres above reference datum) in the mangrove swamps fringing Coral Creek at Hinchinbrook Island, (b) map of Hinchinbrook Island and its fringing mangrove swamps, and (c) bathymetry (in metres) and predicted steady circulation driven by the East Australian Current in Bowling Green Bay. (Adapted from Wolanski and Ridd, 1990.)

where A is the longitudinal eddy diffusion coefficient in the stream, ε is the ratio of the volume in the lateral embayment to that in the estuary, U_0 is the peak tidal current, k^{-1} is the characteristic exchange time between the embayment and the estuary, and σ is the tidal frequency. Most tidal creeks and estuaries in the Great Barrier Reef are mangrove-fringed, and the mangroves are lateral embayments spread all along the estuaries and mangrove creeks. Coral Creek (18.3°S) is a typical mangrove creek (Figure 2.11a). It is about 5 km long with fairly steep banks, and a maximum depth decreasing from 5 m at low tide at the mouth to less than a metre at low tide in the upper reaches of the creek. The creek is surrounded by mangrove swamps, with a gently sloping substrate whose elevation near the creek of about 0.3 m above mean sea level increases to an elevation of about 1.3 m in the far reaches of the swamps. In some places, such as in most bays along the coast of the Great Barrier Reef, many creeks exist one next to the other and form a vast swamp. These intertidal wetlands can be enormous, each covering up to 167 km^2. This is the case, for instance, at Hinchinbrook Channel (18.4°S; Figure 2.11b). In other cases, the mangrove swamp is of limited extent but is backed by a vast intertidal salt flat, such as in Bowling Green Bay (19.4°S; Figure 2.11c). Wolanski and Ridd (1986) modeled analytically the lateral trapping mechanism in mangrove creeks by removing two assumptions in the Okubo (1973) model that are invalid in mangroves, namely, that the depth of water in the creek and in the swamp are equal and that the rate of exchange of water between the creek and the swamp is constant in time. The revised model yields

$$B = \frac{A}{1+\varepsilon} + \varepsilon \; U_0^2 a^2 T/48(1+\varepsilon) \tag{2.3}$$

where (1 – a) is the fraction of the time the swamp has no surface water and 2T is the tidal period. From Equation 2.3, B = 10 to 40 $m^2 s^{-1}$ for small mangrove-fringed creeks such as Coral Creek, and B = 50 to 150 $m^2 s^{-1}$ for Hinchinbrook Channel. This is a mean value for the mangrove-fringed tidal channels. In fact, in their upper reaches, the value of B is smaller since this is where the tidal currents are the smallest (Ridd et al., 1990).

Longitudinal mixing due to vertical mixing and vertical shear dispersion can be parameterized by a longitudinal mixing coefficient K

$$K = bU_0H \tag{2.4}$$

where H is the depth and b is a constant within the range 0.03 to 0.3 (Fischer et al., 1979). Equation 2.4 yields K = 0.05 to 0.5 $m^2 s^{-1}$ for typical mangrove creeks such as Coral Creek. Hence B >> K by one to two orders of magnitude (Wolanski, 1992a), a result implying that lateral trapping is a dominant dispersion mechanism in mangrove swamps.

There is also field evidence to support these predictions. This evidence is twofold. First, in the dry season in small mangrove creeks such as Coral Creek, evapotranspiration results in the salinity increasing along the creek with distance from the mouth (Wolanski, 1992b). Strong tidal currents maintain vertical homogeneity. The salt accumulating in the system as a result of evapotranspiration is balanced at a steady state by salt export from the system. The export is by tidal diffusion. In consequence,

$$A_t E_t S = BA_c \, dS/dx \tag{2.5}$$

where S is the salinity, x the distance from the mouth along the tidal channel, A_t the surface area of the swamp, T_t the evapotranspiration rate, and A_c the stream cross-sectional area. From observed dry-season values of S, B can be calculated from Equation 2.5, giving values for B = 20 to 50 $m^2 s^{-1}$ (Wattaykorn et al., 1990), a range that matches well the predictions from Equation 2.3. The second field evidence pointing to the importance of the lateral trapping phenomenon comes from observations of the flushing of brackish water from Hinchinbrook Channel following a short-lived Herbert River flood (Wolanski et al., 1990). The volume of freshwater remaining in the estuary over a 3-month period following the river flood decreased exponentially with an e-folding constant of about 54 d. The lateral trapping model predicts a residence time of order

$$T = L^2/B \tag{2.6}$$

where L is the characteristic length of the estuary, and gives a value T = 50 d, in agreement with observations.

The trapping of brackish water following a river flood is also enhanced by buoyancy effects in mangrove swamps. Brackish water is pushed back into the mangroves at flood ride, thereby avoiding mixing with more saline water flooding the estuary with the tide. At ebb tide, the trapped water returns in the estuary but forms a plume along the banks of the estuary (Figure 2.12).

2.3 SEDIMENT INFLOW

2.3.1 Riverine Sediment

The Burdekin River is the only large stream for which meaningful sediment discharge measurements have been made (Belperio, 1979, 1983). The peak suspended solids concentration reaches 0.5 g l^{-1} during floods, although com-

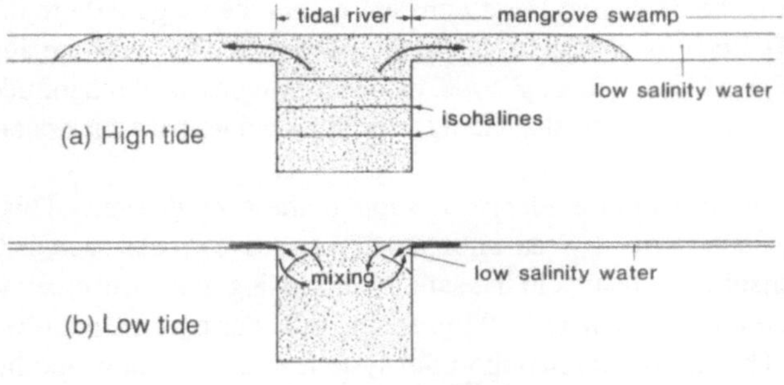

FIGURE 2.12. Cross-channel distribution of salinity in a mangrove-fringed estuary following a river flood. (From Wolanski and Ridd, 1986. Reproduced with the kind permission of Academic Press.)

monly the concentrations are smaller. The silt and clay size sediment input by the Burdekin River to the Great Barrier Reef is about 3 million t $year^{-1}$. Most of this sediment may be discharged in one day during a major flood. Extrapolated to other catchments along the Queensland coast, the riverine sediment input to the Great Barrier Reef is about 28 million t $year^{-1}$. Assuming this sediment stays trapped along the coast, the inflow corresponds to a horizontal coastal progradation of about 1 to 1.5 m $year^{-1}$, accompanied by a vertical accumulation in intertidal environments of up to 0.08 m $year^{-1}$.

In smaller rivers, the suspended sediment concentration can vary several orders of magnitude in a few hours following the rapid rise and fall of the water discharge after intense tropical storms (Figure 2.7a). The highest concentrations are found in the rising stage of the river flood. The waters are extremely turbid, with a suspended solids concentration exceeding 1.5 g l^{-1} , This high turbidity is in this case probably due to erosion from a derelict mine site (Hart et al., 1988). A high turbidity in rivers along the Queensland coast is quite common in all areas where man has removed the protective vegetation for farming, road work, mining, building sites, and other land clearing, because of the erosion caused by intense rainfall in tropical areas (Cameron McNamara, 1985; Piorewicz and Ryall, 1991). In primary rain forest, the river waters are, by comparison, commonly quite clear and the suspended sediment load is minimal.

2.3.2 Estuarine Trapping

The Burdekin River sediment reaches the coast in river floods, since at such times the estuarine waters are fresh almost to the river mouth. This is not the case, however, in most other estuaries which act as sediment traps or filters. This effect has only been studied in detail in the Great Barrier Reef in the Normanby River (14.5°S). This estuary is particularly turbid, with a suspended sediment concentration reaching 14 g l^{-1} near the bottom. Three current metres and a string of six optical backscatter nephelometers were deployed at point A (Figure 2.13a) in the estuary. The current metres were deployed, in 3.8-m depth at low tide, at 1.2, 2.2, and 3.3 m above the bottom. The nephelometers were deployed from 0.5 to 3 m above the bottom. Figure 2.13b shows over two tidal cycles the fluctuations of sea level, water velocity, and suspended sediment concentration (SSC). The vertical shear of the currents was negligible in the bulk of the water column away from the lower 1 m where no data are available. The flood tidal currents were slightly larger than the ebb ones. The SSC at all elevations was largest at peak flood (event B1) and ebb tidal currents (event A1), because of erosion and vertical mixing of the fine sediment. However, at the bottom of the string, high SSC values were also found at slack low water (event C1), while the upper waters became clear. In contrast, at slack high water, SSC decreased throughout the water column (event D1). This pattern may be attributed to the tidal asymmetry; the change from ebb to flood tidal currents occurs more rapidly than the change from flood to ebb tide; hence, at slack high tide, but not at slack low tide, the turbulence has time to dissipate. Because the settling velocity of the suspended sediment is dependent on the turbulence, the suspended sediment cannot settle out of the bottom turbid layer

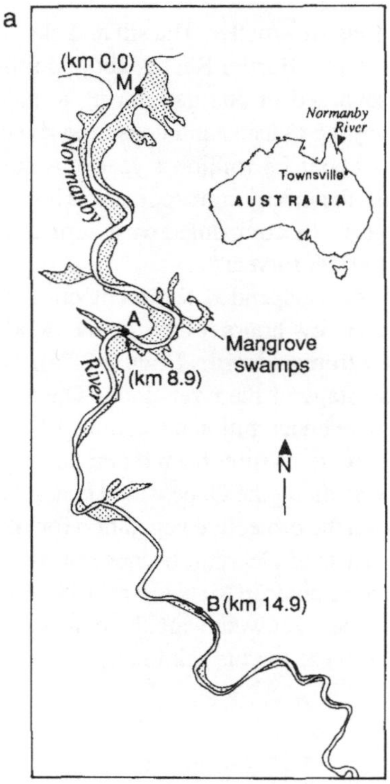

FIGURE 2.13. (a) Location map of the Normanby River and (b) time series plot (in day no. in September 1989) of the sea level at site M, water currents at three elevations at site A, and suspended sediment concentration at site A. Events A1, B1, C1, and D1 are discussed in the text. (From Wolanski et al., 1992a. Reproduced with the kind permission of *Journal of Coastal Research*.)

at slack low tide. This asymmetry, as well as that of the tidal currents, generates a net up-river transport of fine sediment in the Normanby River estuary, and hence long-term trapping of the sediment (Wolanski et al., 1992a).

Tidal currents in the Fly River estuary (8.5°S; Figure 2.14a) are also asymmetric. The estuary is macrotidal and only weakly stratified in salinity. The estuary is also very turbid and a near-bottom fluid-mud layer is found in the turbidity maximum zone where the suspended solid concentration can reach 100 g l^{-1} (Wolanski and Eagle, 1991). The settling velocity w_f of the suspended sediment varies nonlinearly with the SSC (Figure 2.14b). Flocculation settling prevails for SSC <2 g l^{-1}, w_f increasing with increasing values of SSC. Hindered settling occurs for SSC >6 g l^{-1} when w_f decreases with increasing values of SSC. Such a dependence of w_f on SSC is a characteristic of cohesive sediments (Ross and Mehta, 1989).

To examine the nature of the sediment, microphotographs were obtained in the Fly River estuary. In freshwater, the sediment is largely uncoagulated (Figure 2.15a) and composed of silt and clay particles. In brackish water (Figure 2.15b), the fine sediment is coagulated. The floc size has a median value of about 30 μm

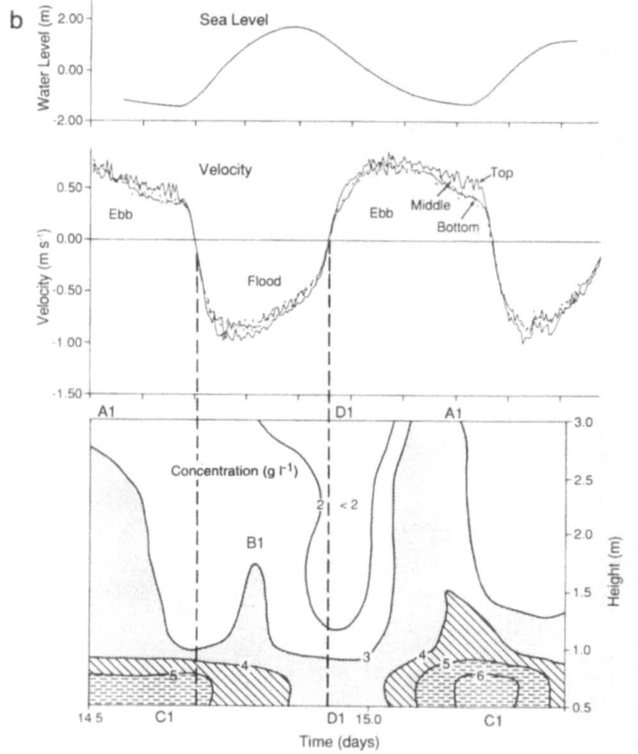

FIGURE 2.13 (continued).

although on occasion large flocs of 200 μm are observed. The SSC varies considerably with the tides, with a succession of events of deposition, erosion, and vertical entrainment of sediment. Following slack tidal currents, the sediment settles out, with a large time lag, from the bulk of the water column except from the near-bottom fluid mud layer (Wolanski et al., 1992c).

The key processes controlling the fate of fine sediments in the Fly River estuary are sketched in Figure 2.16. The fine sediment is uncoagulated in the freshwater part of the estuary. At a salinity near 1 to 4 psu, the fine suspended sediment coagulates and its settling velocity increases; it thus remains preferentially near the bottom. Some of the fine sediment escapes as small flocs and moves downstream; this sediment is reentrained upstream with the up-river, near-bottom baroclinic estuarine circulation. In addition, the tidal asymmetry of the currents results in an up-river pumping of sediment. It results in a turbidity maximum zone located slightly downstream of the salinity intrusion limit. Thus, the Fly River estuary is a sediment trap. However, some sediment is still exported in suspension as small flocs during ebb tides, although the determination of this net flux is difficult as the estuarine return coefficient is >0.9, i.e., at least 90% of the water that leaves the estuary at ebb tides returns at the following ebb tides.

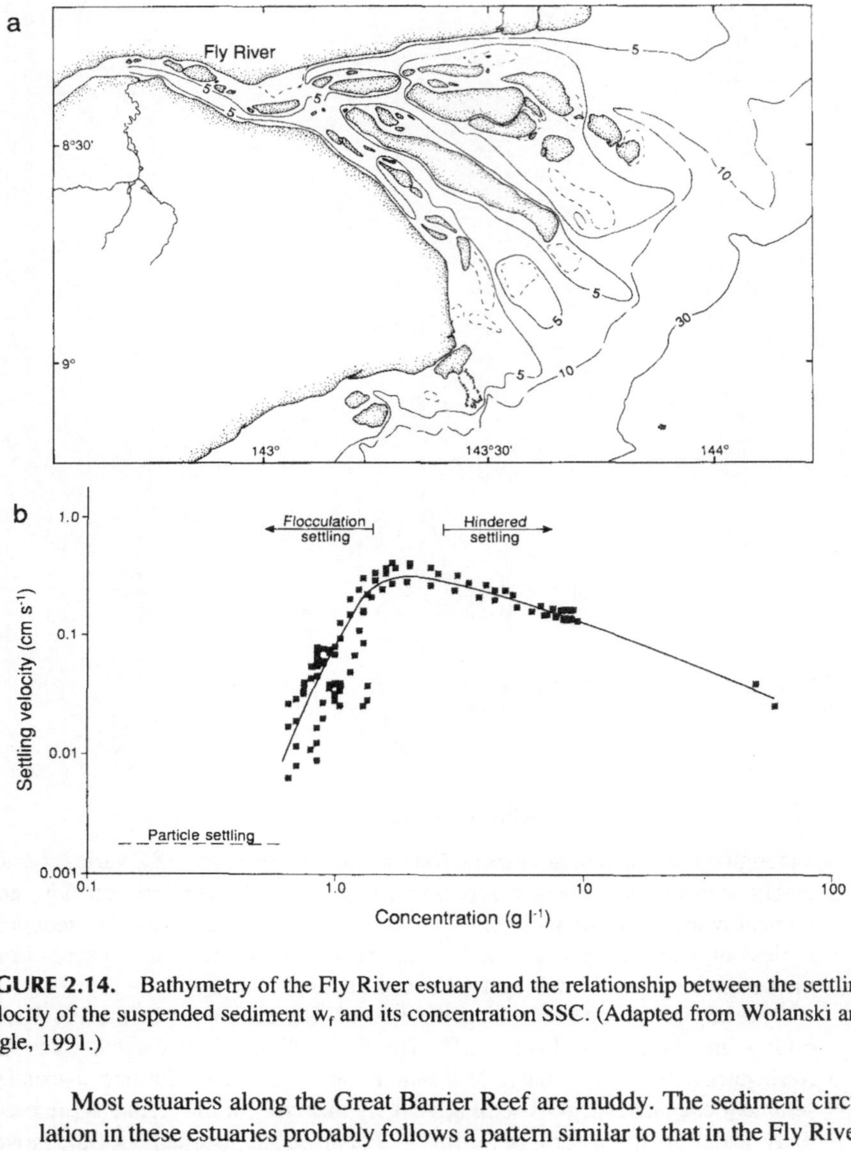

FIGURE 2.14. Bathymetry of the Fly River estuary and the relationship between the settling velocity of the suspended sediment w_f and its concentration SSC. (Adapted from Wolanski and Eagle, 1991.)

Most estuaries along the Great Barrier Reef are muddy. The sediment circulation in these estuaries probably follows a pattern similar to that in the Fly River, but has not been studied.

2.3.3 Dredging

Dredging of navigation channels also provides to coastal waters an important (at a local scale) source of cohesive sediment. Dredging exposes old sediment to erosion and releases sediment into the water column. Townsville Harbour is accessible by a navigation channel, the Platypus Channel, dredged through Cleve-

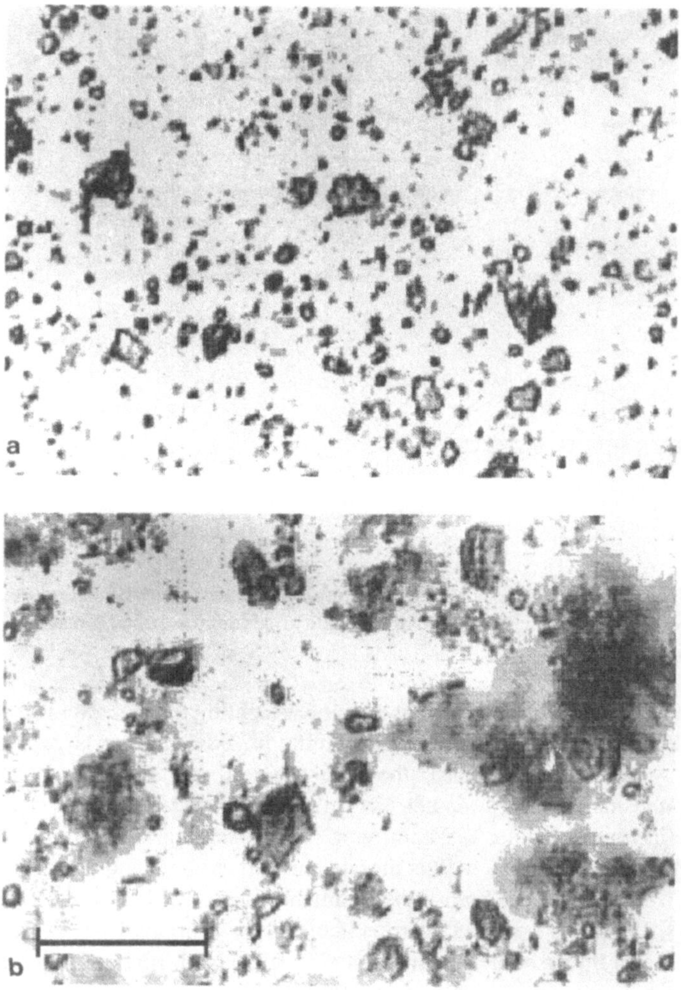

FIGURE 2.15. Microphotographs of suspended sediment in the Fly River estuary in (a) fresh water and (b) brackish water. The bar represents 40 μm.

land Bay (19.3°S; Figure 2.17). The channel is dredged to typically 6 m below the natural sea floor into indurated Pleistocene clay and sand (Johnson and Carter, 1987). Fine sediment is mixed into the water column during dredging as a result of mechanical disturbance of the bed and the overflow from the storage tanks, and from dumping at sea at the offshore dump site. The dredge spoil is dumped when the ship opens trap doors while underway. A turbid wake is visible. Successive acoustic transects in the wake in calm weather show the turbid water settling out initially very rapidly (in a few minutes of 10-m depth) as a negatively buoyant plume forming a near-bottom turbid plume (Wolanski et al., 1992b). Thereafter,

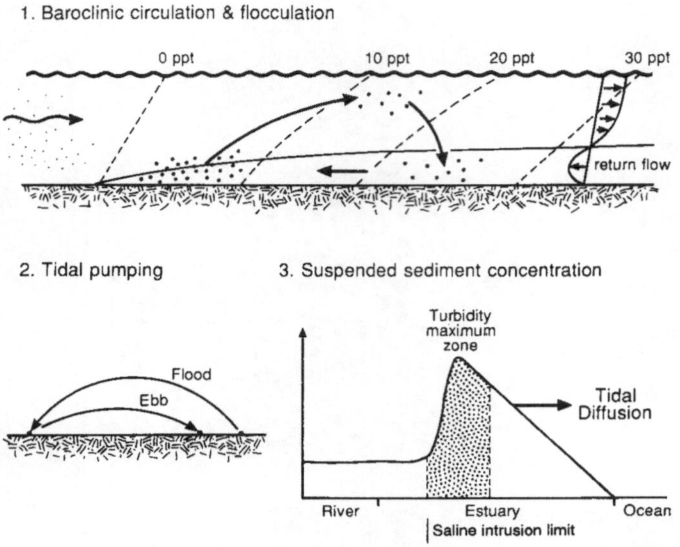

FIGURE 2.16. Sketch of the key processes controlling the transport of fine sediment in the Fly River estuary and the resulting formation of a turbidity maximum zone. (Adapted from Wolanski et al., 1992d.)

clearing of this layer is slower as it involves settling of the individual mud flocs. Profiles of suspended sediment concentration (SSC) in that layer show that in calm weather (Figure 2.17b) the flocs have all but settled out 15 min after dumping. In moderate weather (15 to 20-knot wind), settling is much slower, while in rough weather (25-knot wind) no net settling is apparent 15 min after dumping. The sediment is thus mobile and can escape the dump site (Wolanski et al., 1992b). Further, even if the sediment reaches the bottom in calm weather, it may still be resuspended later, presumably by pumping by long waves. Some of the spoil thus is widely redistributed in Cleveland, Bay, a finding also borne out from geomorphological studies (Pringle, 1989).

Sediment is also released in the water column by the dredger itself, as a result of the overflow, disturbance of the bed, and ship-generated turbulence. Wolanski and Gibbs (1992) used profiling nephelometers and high-frequency echo soundings to obtain data on the fate of the sediment in the platypus Channel after passage of the dredger. They found a negatively buoyant plume extending initially all the way to the surface and settling at a speed of about 0.8 cm s^{-1}. In windy conditions a fraction of this sediment escapes from the channel and drifts away with the prevailing wind-driven currents.

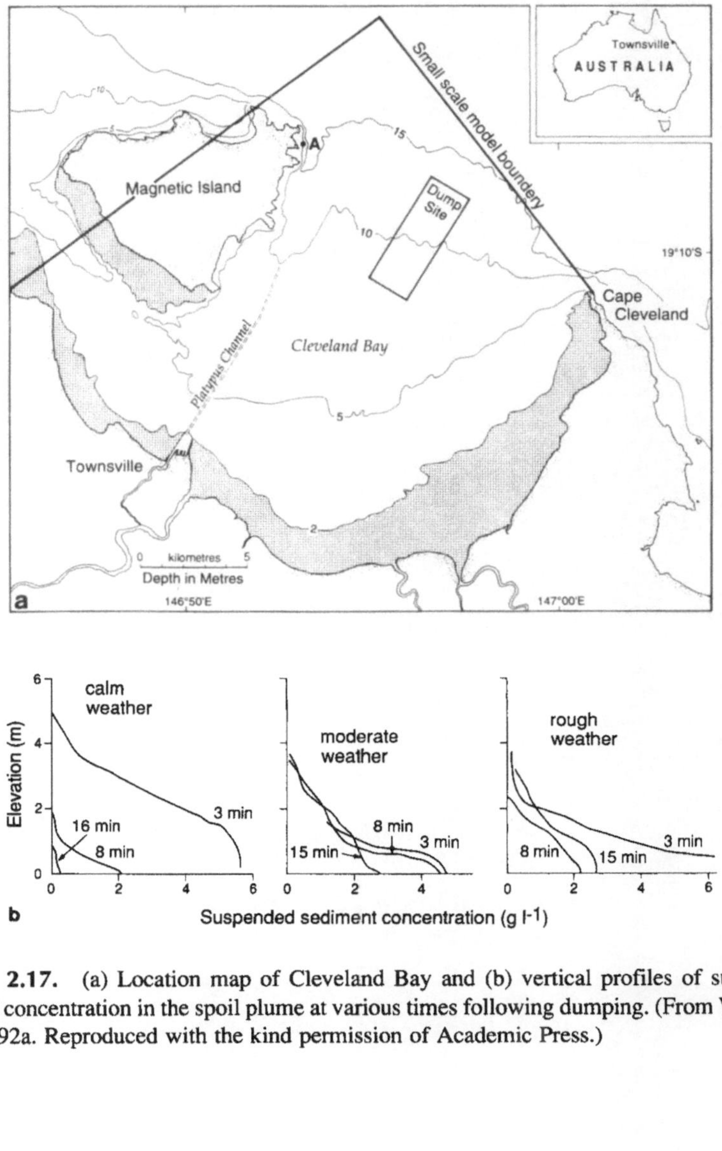

FIGURE 2.17. (a) Location map of Cleveland Bay and (b) vertical profiles of suspended sediment concentration in the spoil plume at various times following dumping. (From Wolanski et al., 1992a. Reproduced with the kind permission of Academic Press.)

3 Water Properties and Patchiness

The properties of the Great Barrier Reef water show a gradient, sometimes smooth and continuous, sometimes discontinuous and step-like, between those of coastal waters and those of the Coral Sea. In contrast, however, to other reef-free continental shelves, the coral reefs can also create their own water characteristics both within lagoons and around the reefs, particularly in areas where reef density is high.

3.1 TYPES OF WATER

Three types of water masses exist throughout the year in the far northern Great Barrier Reef and Torres Strait. In the far-northern Torres Strait, shallow waters and strong tidal currents maintain vertical homogeneity in temperature and salinity (Figure 3.1a). In the Gulf of Papua (8.5°S), the Fly, Purari, and Kikori Rivers and other coastal streams have a combined freshwater discharge of 15,000 $m^3 s^{-1}$. This outflow creates a brackish water plume in the top 20 m of the Gulf of Papua. In calm weather, this surface layer is continuously stratified in salinity (Figure 3.1b). During strong winds, wind mixing results in a well-mixed brackish surface layer, separated by a sharp (1- to 2-m thick interface from the clear, saline waters below (Wolanski et al., 1992c). Finally, in the Coral Sea, a uniform salinity is found in the top 100 m, while a temperature stratification may exist; in summer, when winds are light and heating is maximum, the waters are continuously stratified in temperature, although a number of step structures may be present (Figure 3.1c); in winter, when the southeast winds prevail, the top 120 m can be well mixed (Figure 3.2).

Below a depth of 100 m, the salinity increases to a salinity maximum typically located 150 m below the surface (Figure 3.2). These T-S properties are fairly typical of the Coral Sea along the entire length of the Great Barrier Reef. The mixed-layer depth is much larger than the water depth over the continental shelf; thus, subthermocline water is only found on the shelf following upwelling events.

3.2 CORAL SEA NUTRIENTS

While surface waters are generally oligotrophic, as typical of coral reefs, water masses below the thermocline are nutrient rich. Nutrient data of Andrews and

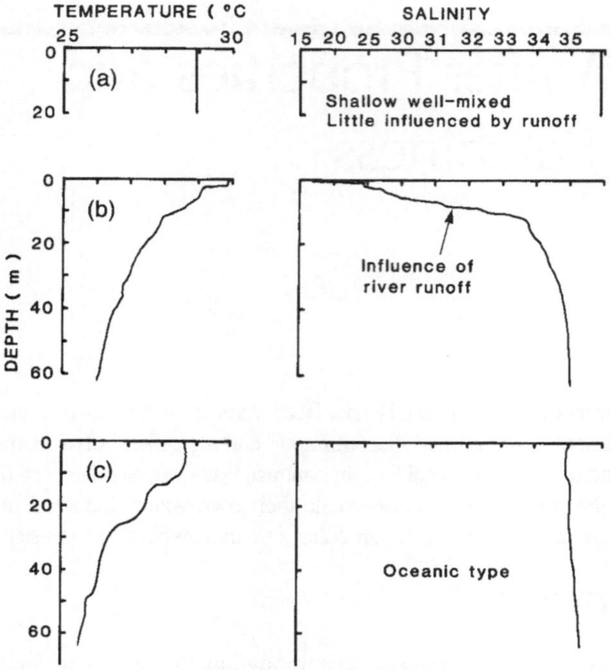

FIGURE 3.1. Typical vertical profiles of temperature and salinity in (a) Torres Strait, (b) coastal waters of the Gulf of Papua, and (c) at the shelf break of the Gulf of Papua. (Adapted from Wolanski et al., 1984b).

Gentien (1982) and Tomczak (1983) suggest that, in these deep waters for temperature T between 10 and 25°C, the concentration C of nutrients approximately follows the equation

$$C = a - b\,T \tag{3.1}$$

where C is expressed in millimoles per cubic metre and a and b are constants (for nitrate, a = 30.2, b = 1.2; for phosphate, a = 2.2, b = 0.09; for silicate, a = 17.6, b = 0.8). Equation 3.1 shows that the nutrient concentrations decrease with increasing temperature, i.e., increase with depth below the thermocline. Upwelling events may thus be an important source of nutrients for the Great Barrier Reef continental shelf, in the same way that upwelling is important as a nutrient source in most other shelf regions worldwide.

3.3 SHELF WATERS

Only in the central region has there been an extensive study of water properties on the mid- and outer shelf and the adjoining Coral Sea (Andrews, 1983a, b;

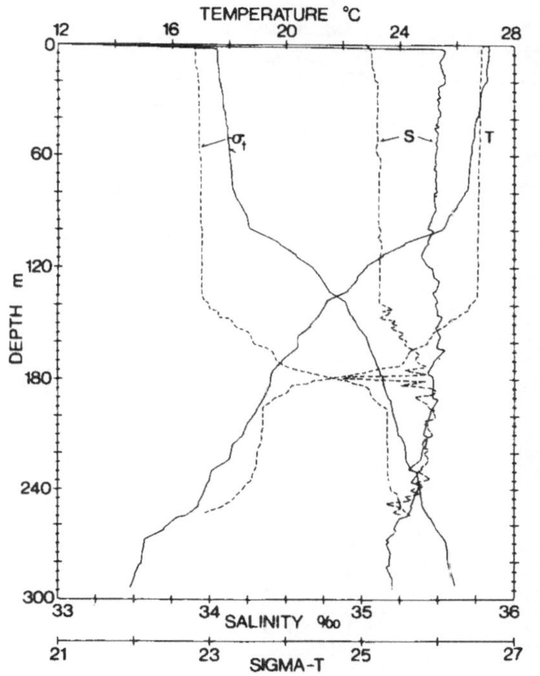

FIGURE 3.2. Profiles of temperature (T), salinity (S), and density (σ_t) in the Coral Sea near Raine Island. Solid lines: November 16, 1981; dashed lines: May 1, 1982. (From Thomson and Wolanski, 1984.)

Bellamy et al., 1982; Andrews and Gentien, 1982). From these data, four water masses were identified (Figure 3.3): the subtropical lower water, the shelf-break water, the Coral Sea surface water, and the shelf water. Shelf waters were found over the shelf for water depths <70 m. Shelf-break water was found in the bottom 30 m on the continental slope where the total depth was between 70 and 200 m. Well-mixed conditions prevailed in the Coral Sea surface water, which comprises the top 65 m of the water column where the water depth is >300 m.

The temperature and salinity fluctuate seasonally. The amplitude of the variation is largest on the inner shelf, particularly for salinity, which is influenced by river runoff. The shelf waters are colder in winter (southeast trade wind season). The seasonal variations are largely confined to the surface waters above the Subtropical Lower Waters, which have a typical temperature and salinity of 21.5°C and 35.7 psu, respectively (Andrews, 1983a).

Seasonal fluctuations in water temperature on the shelf are large, up to 11.5°C in embayments such as Cleveland Bay (19.3°S; Walker, 1981b). Smaller fluctuations are experienced offshore. The longest temperature time series published spans 2.5 years, from 1980 to 1982 at a mooring site near Green Island (16.8°S), and is shown in Figure 3.4 together with the temperature data from mooring sites

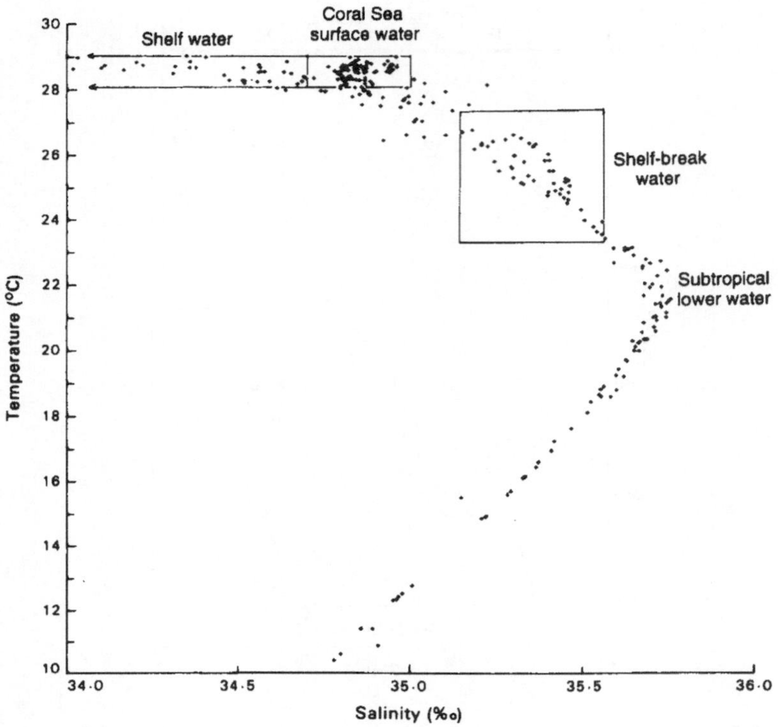

FIGURE 3.3. Temperature-salinity diagram for 0 to 200 m in the wet season of 1981 on the shelf and the slope of the central region. (Adapted from Andrews, 1983a.)

also on the mid-shelf near Linnet Reef (14.8°S) and Lizard Island (14.6°S), on the outer shelf at Euston Reef (16.7°S), and on the inner shelf at Cape Upstart (19.7°S). These sites are up to 600 km apart. The seasonal temperature fluctuations are the dominant signal. Also apparent are interannual fluctuations, as well as low-frequency fluctuations of 5 to 30 d which are incoherent from site to site. There are also variable along-shelf and cross-shelf temperature gradients. There are only small temperature fluctuations at tidal periods. Some low-frequency temperature fluctuations are due to the upwelling of shelf slope water onto the shelf. Andrews and Gentien (1982) found evidence of such intrusions penetrating two thirds of the way from the shelf break toward the coast.

Weekly measurements for a year of temperature and salinity on a transect from the coast to Keeper Reef (18.7°S; Figure 1.5) showed that the waters are well mixed during the southeast trade wind season (Wolanski et al., 1981). Under very dry conditions, from September to November, coastal waters can become, by evaporation, more saline (by 0.5 psu) than, and sink under, mid-shelf waters (Figure 3.5). The extent of this mechanism is not known (Walker, 1982).

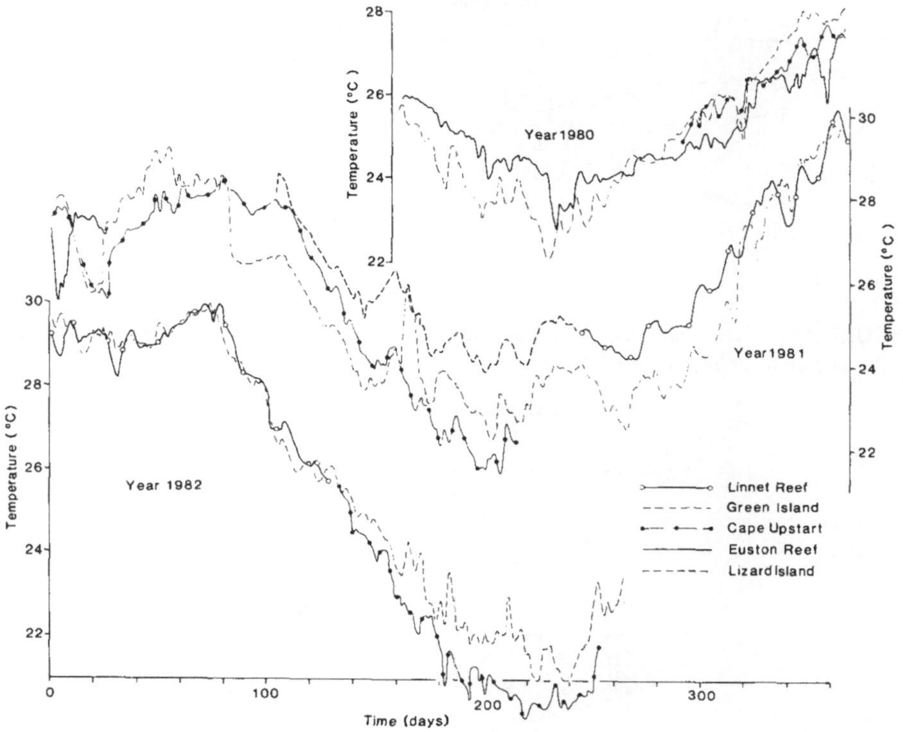

FIGURE 3.4. Time series of the low-frequency temperature at various current-metre sites. (From Wolanski and Pickard, 1985. Reproduced with the kind permission of Academic Press.)

3.4 RIVER PLUMES

In summer, following rainfall and runoff, coastal waters are stratified. This is apparent in the salinity distribution in a cross-shelf transect (Figure 3.6). The transect intersects the Burdekin River plume. The salinity fluctuates widely from week to week, in response to the fluctuations in river runoff. Only during peak river floods does the Burdekin River plume touch the bottom.

Wolanski and van Senden (1983) used four ships to survey, at weekly intervals for 3 weeks, the salinity distribution along a 400-km stretch of the continental shelf between the mouth of the Burdekin River and Cairns in January and February 1981. The Burdekin River discharge started on January 15, 1981, peaked at about 13,000 $m^3 s^{-1}$ on January 23, and decreased to about 2000 $m^3 s^{-1}$ on January 28 (Figure 2.7b). The distribution of surface salinity (Figure 3.7) shows that the river plume stayed close to the coast and moved north. Significant changes in salinity at the Great Barrier Reef occurred only north of Britomart Reef (18.3°S). Current-metre data from Wolanski and van Senden (1983) showed that the

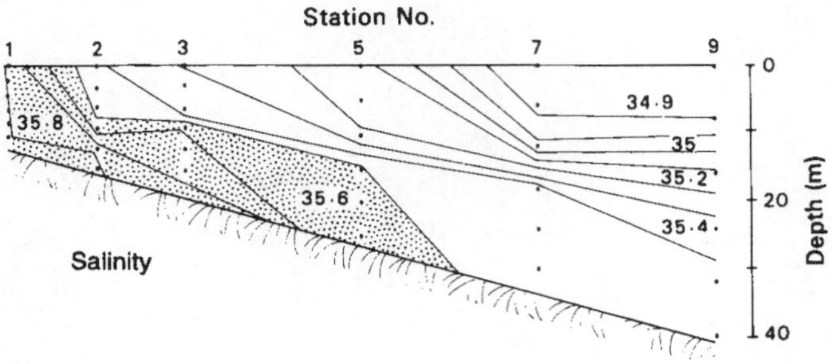

FIGURE 3.5. Cross-shelf distribution of salinity along the Keeper Reef transect, November 1991. Location site in Figure 1.5.

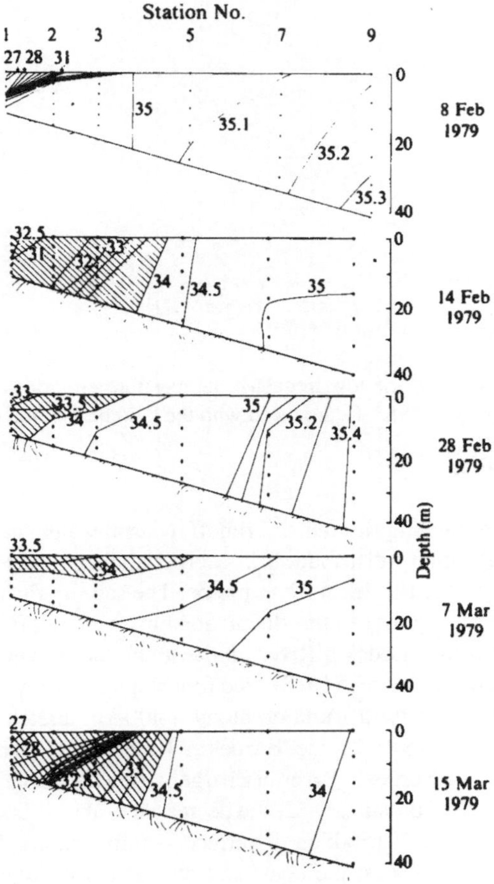

FIGURE 3.6. Cross-shelf distribution of salinity, Keeper Reef transect, during the wet season of 1979. Location site in Figure 1.5. (Adapted from Wolanski and Jones, 1981.)

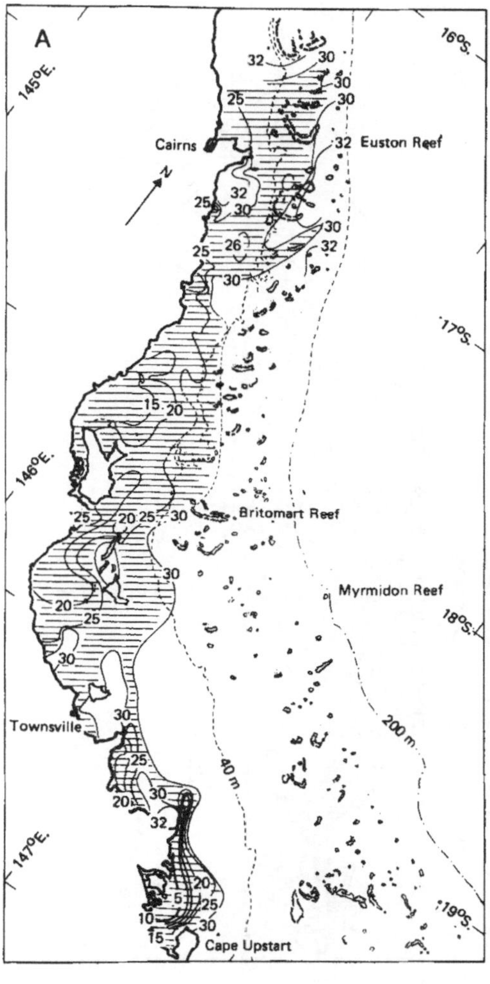

FIGURE 3.7. Surface salinity (a) 2 d, (b) 8 d, and (c) 16 d after a major freshet from the Burdekin River on January 18 to 19, 1981. Discharge data are shown in Figure 2.7b. (Adapted from Wolanski and van Senden, 1983.)

northward movement of brackish waters was opposite that of the southward currents (0.05 to 0.1 m s^{-1}) on the mid-shelf off Cape Upstart (19.7°S).

What is also apparent in Figure 3.7 is the breakdown of the plume into patches. More details of such patches were obtained by Wolanski and Jones (1981) following the Burdekin River flood in January 1980. They found that the brackish water patches formed a few days after cessation of the river flood. The patches were preferentially located near embayments and headlands (Figure 3.8). The residence time is longest on the order of a few weeks, and the flushing rate

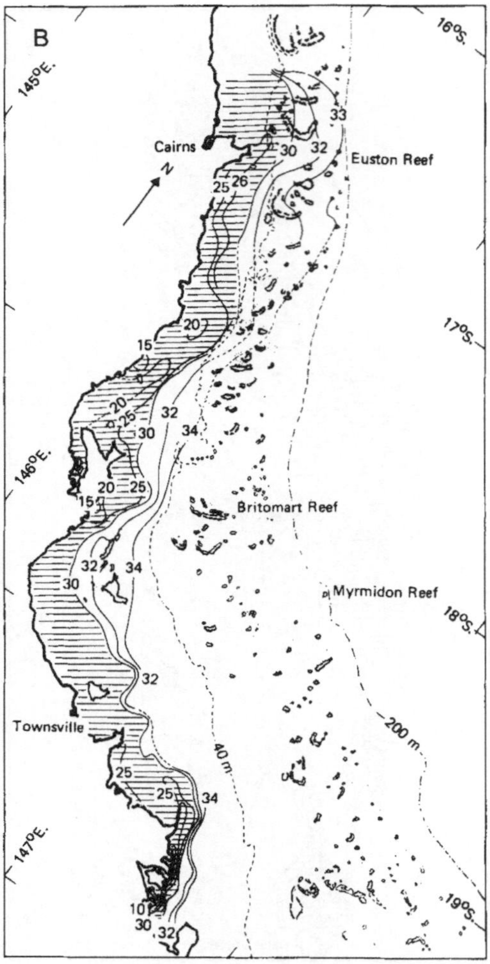

FIGURE 3.7 (continued).

smallest, in these embayments. The injection of pulses of Burdekin River plume water in Cleveland Bay (19.3°S; Figure 2.17) generates salinity fluctuations, measured at weekly intervals, of up to 8 psu (Walker, 1981a).

These observations imply that during a flood of the Burdekin River, the plume moves northward as a geostrophic, density current, with the coast on its left in the Southern Hemisphere. Laboratory experiments (Stern, 1980) suggest that the width L of this current and the speed c of the front are given by

$$L = 0.42\sqrt{gh\frac{\Delta\rho}{\rho}}\ f^{-1} \tag{3.2}$$

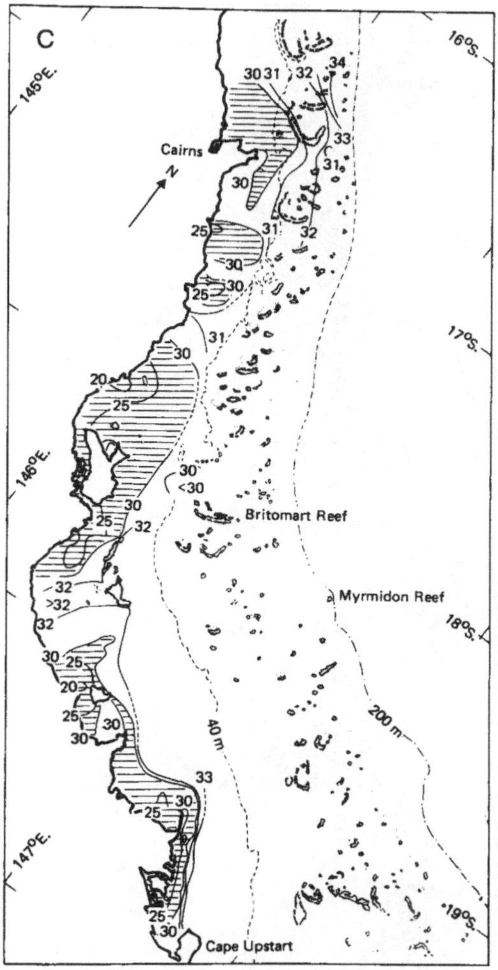

FIGURE 3.7 (continued).

$$c = 1.1\sqrt{gh\frac{\Delta\rho}{\rho}} \tag{3.3}$$

where h is the depth of the plume, f ($= -4 \times 10^{-5}\ s^{-1}$) the Coriolis parameter, and $\Delta\rho$ the difference in density ρ between river plume and receiving waters. The field data north of Townsville allow estimates of h and $\Delta\rho/\rho$ and from such estimates, Equations 3.2 and 3.3 predict L = 12 km and c = 85 km d^{-1}.

If the river plume does not touch the sea floor, bottom friction effects presumably may be neglected in the calculation of the northward water current speed v in the plume from the depth-averaged thermal wind equation

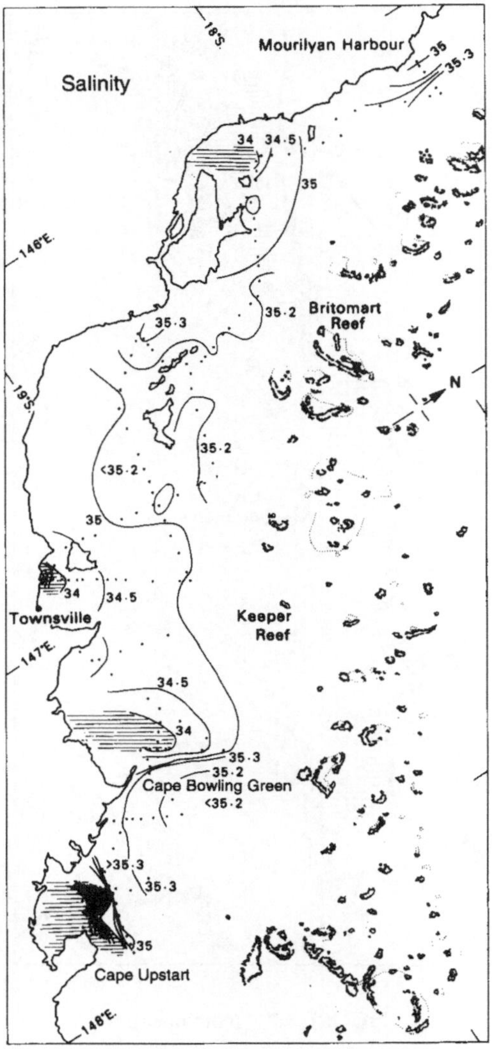

FIGURE 3.8. Surface salinity on January 15 to 16, 1980. (Adapted from Wolanski and Jones, 1981.)

$$fv = \frac{g}{2\rho}\frac{\partial \rho}{\partial x} \tag{3.4}$$

where x is the axis-oriented cross-shelf. From Equation 3.4, v = 0.18 m s^{-1} north of Townsville, a result in agreement with the current-metre observations of Wolanski and van Senden (1983). However, in large floods the Burdekin River plume touches the bottom (Figure 3.6), and bottom friction effects become important. The river flood then raises the coastal sea level (Csanady, 1978, 1982).

The sea surface slopes downward to offshore, and this generates a geostrophic, northward, longshore current of 0.05 to 0.1 m s^{-1}. This current in turn generates a longshore sea-level gradient. The buoyancy inhibits mixing between river plume and receiving waters.

The Burdekin River plume dynamics are qualitatively similar to the situation along the Nicaragua coast (11 to 15°N), where brackish water flows longshore for 400 km along the coast, remaining within 20 km of the shore and not reaching the coral reefs further offshore (Murray and Young, 1985). The Burdekin River plume width increases faster with distance from the source than the Nicaraguan River plume, presumably because the rugged topography of the Great Barrier Reef enhances cross-shelf mixing. This mixing is in addition to that associated with unsteady, coastal, baroclinic currents (Stern, 1980; Nof, 1981). This enhanced mixing is evident in Figure 3.8, where the salinity contours off Cape Bowling Green are oriented cross-shelf.

Brackish water can be detrimental to coral reefs. An important parameter is the duration and dilution of the intrusion of brackish water plumes on the coral reefs. Salinity data at Britomart Reef (18.3°S) during the 1981 Burdekin River flood show small tidal fluctuations but large low-frequency fluctuations (e.g., events A and B) associated with the passage of brackish water patches (Figure 3.9). These patches are not uniform in salinity. Imbedded within them are very low salinity patches or "bubbles" (e.g., event C). Direct rainfall following the passage of a tropical cyclone was responsible for the low-salinity spike (event D, corresponding to a salinity decrease of about 2.4 psu). There are few other observations of the effect of direct rainfall on the reef. A salinity decrease of 0.8 psu at Keeper Reef (18.7°S) following the passage of a tropical cyclone has been reported by Wolanski et al. (1981).

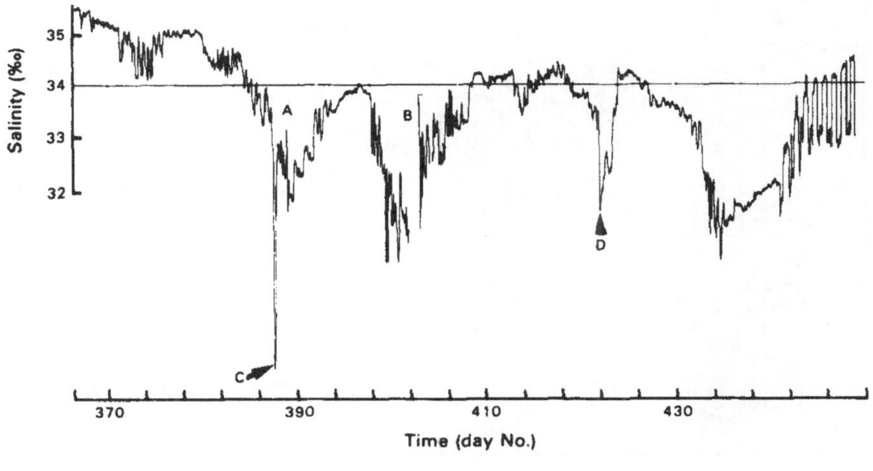

FIGURE 3.9. Time series at 10-min intervals of the salinity 10 m below the surface in the western entrance of Britomart Reef. Time is expressed as day no. in 1981. Events A to D are discussed in the text. (From Wolanski and van Senden, 1983. Reproduced with the kind permission of CSIRO Editorial Services.)

Large temporal variations of salinity were found in the wet season, but the spatial distribution of the brackish water patches was not measured, on the inner shelf in Princess Charlotte Bay (14.3°S) influenced by the Normanby, Bizant, and Kennedy Rivers, and between Cooktown (15.5°S) and Innisfail (17.5°S) influenced by a number of small rivers (Hamilton, unpublished data).

Only in the far northern Torres Strait does the salinity show small seasonal fluctuations (<2 psu; Wolanski et al., 1984b). The surface and 20-m salinity distributions are practically equal in the Torres Strait, because of strong tidal mixing in shallow waters. The brackish water found in the northern Torres Strait (Figure 3.10) originates from the Gulf of Papua. The brackish water intrusion is topographically confined to the Great North East Channel by the topography, being prevented from spreading westward into Torres Strait by the Warrior Reefs, and eastward onto the Great Barrier Reef shelf by a dense matrix of reefs. Gulf of Papua brackish water can also intrude into Torres Strait through Missionary Passage, the narrow passage between Papua New Guinea and the Warrior Reefs (Figure 3.11). This intrusion is apparent in the September 1990 data corresponding to the southeast trade wind season. However, in December 1990, in the northwest wind season, the water was fresher to the east than to the west of the

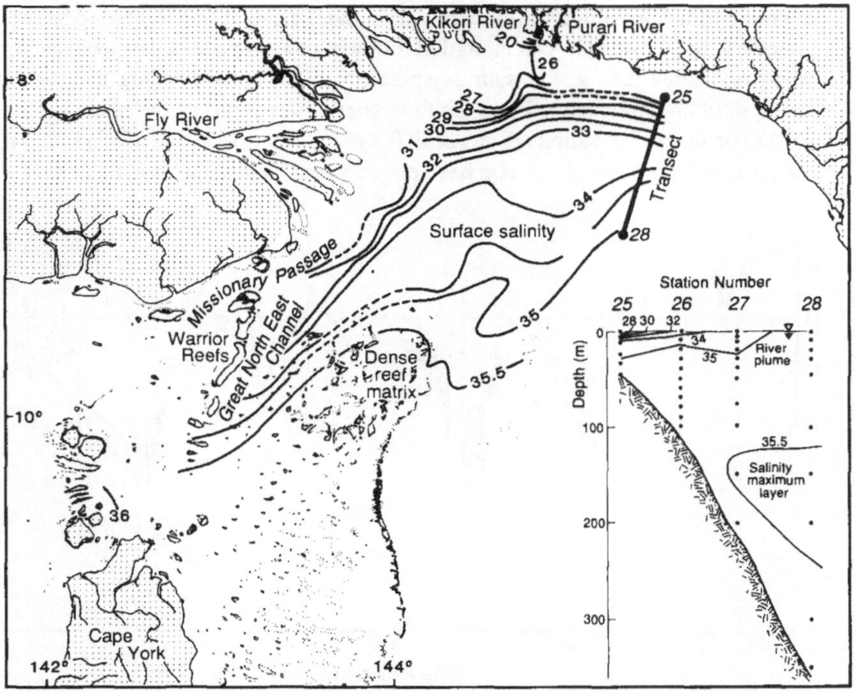

FIGURE 3.10. Surface salinity distribution in November 1979, and cross-shelf distribution of salinity, in the Gulf of Papua and the northern Torres Strait. (Adapted from Wolanski and Ruddick, 1981.)

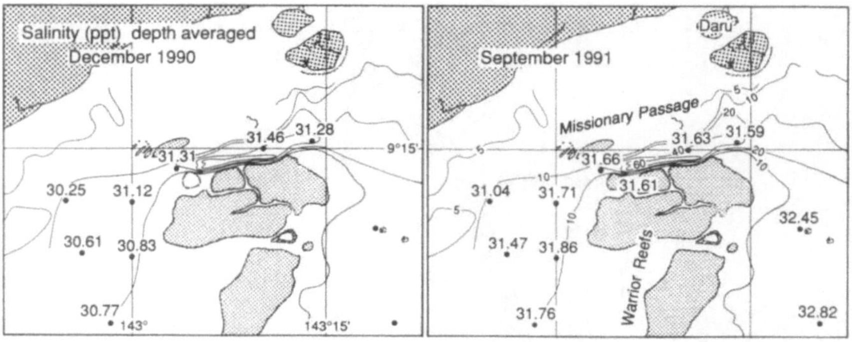

FIGURE 3.11. Depth (in metres) and water salinity (vertical homogeneity prevailed) in the far-northern Torres Strait around Missionary Passage.

Warrior Reefs, a finding suggesting that the waters west of the Warrior Reefs did not originate from the Gulf of Papua, but instead was Irian Jaya coastal water.

The unpredictability, unsteadiness, and patchiness of the river plume intrusion on the Great Barrier Reef makes data collection very difficult. The understanding of river plume dynamics in the Great Barrier Reef remains mostly descriptive. Yet the fate of river plumes is a key question for the management of the Great Barrier Reef because of the inflow of pollutants from human activities on land (see Section 3.6).

3.5 TOPOGRAPHICALLY CONTROLLED PATCHINESS

Most coral reef lagoons have a wide opening, e.g., the one at Bowden Reef (Figure 1.7), enabling active mixing between lagoons and shelf waters. Nevertheless, lagoon waters are generally slightly cooler or warmer, according to prevailing meteorological conditions, than the surrounding shelf waters (Figure 3.12a). In calm weather, top-to-bottom temperature differences in a 20-m-deep lagoon can reach 1°C, while at the same time the thermal stratification can be negligible in the surrounding Great Barrier Reef waters. Eddies in the lee of islands and coral reefs also trap water with a temperature different from that of adjoining waters (Figure 3.12b).

Strong tidal currents through reef passages result in a Bernoulli-type tidal pumping effect (Figure 3.13b), resulting in an upward sucking of deeper, colder water onto the reef passage (Thompson and Golding, 1981). This upwelled water then propagates shoreward, along the bottom, as a tidal jet-vortex pair system (Figure 3.13a). Resulting near-bottom temperature fluctuations at the semidiurnal tidal period can be up to 3°C. The upwelling intensity is strongly related to the topography, particularly the shape of the passage, including the presence or absence of a submarine canyon. The influence of the topography is illustrated in Figure 3.14, which shows the near-bottom temperature signal for several months near two reef passages in the northern Great Barrier Reef. While the seasonal

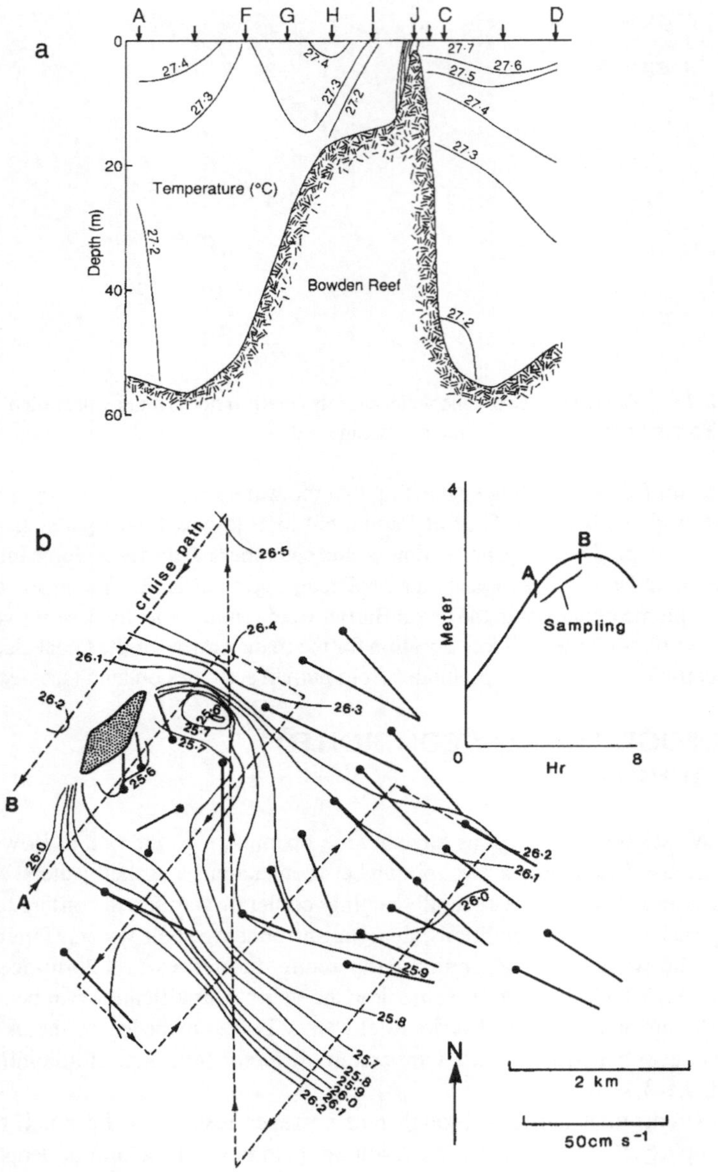

FIGURE 3.12. (a) Cross-reef temperature distribution at Bowden Reef and (b) currents (lines out from dots) and temperature distribution around Rattray Island. (Adapted from Wolanski et al., 1984a, 1989.)

trend was similar at all sites, the semidiurnal temperature fluctuations, hence the upwelling intensity, were much larger at the mooring near Wilson Reef (14°S) than near Franklin Reef (13°S). Tidal currents were of comparable magnitude at

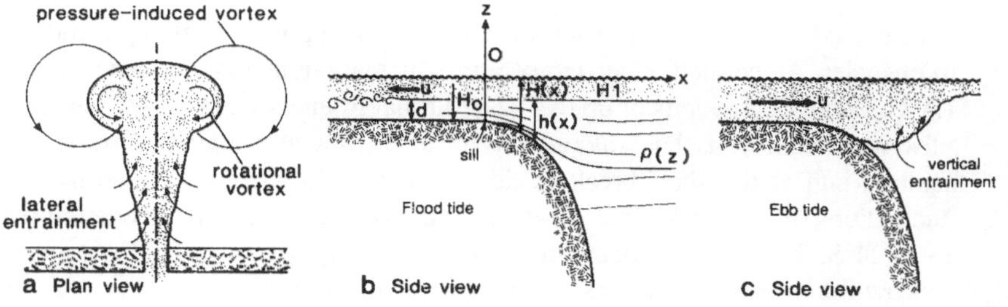

FIGURE 3.13. Sketches of a tidal jet-vortex pair system at flood tide in (a) plan view and (b) side view, and (c) the upper continental slope in side view at ebb tide. (From Wolanski et al., 1988a. Reproduced with the kind permission of Academic Press.)

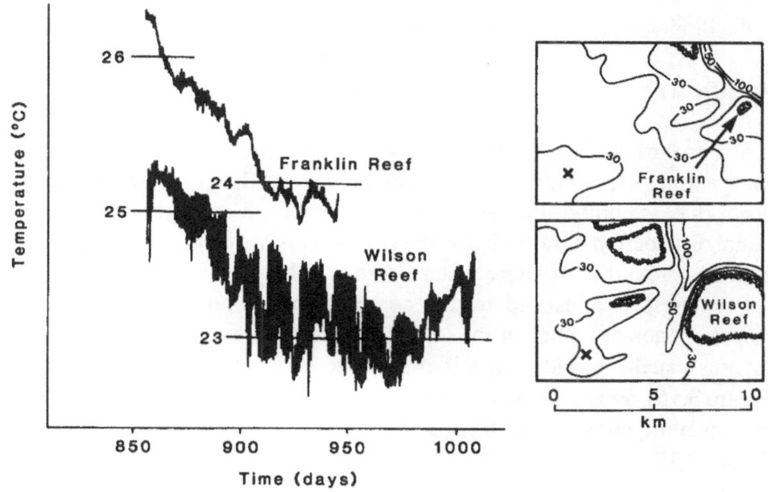

FIGURE 3.14. Time series plot of the near-bottom temperature at mooring sites near two reef passages. Time is expressed as day number in 1980 to 1981. (Adapted from Wolanski, 1986c.)

the three sites, oriented cross-shelf, flood tidal currents oriented shoreward (about westward) and ebb tidal currents oceanward. The temperature profile in the Coral Sea showed a well-mixed surface layer 80 m thick, so that the thermocline was located 50 m below the elevation of the shelf. The shelf waters were vertically well mixed except during an upwelling event. The stronger upwelling events at Wilson Reef can be attributed to the presence of a small submarine canyon facilitating the intrusion of upwelled water onto the shelf.

On the outer shelf near Myrmidon Reef (18.3°S), near-bottom temperature fluctuations of semidiurnal frequency of up to 8°C have been observed and were generated by the enhancement of internal tides by a complex topography (Wolanski and Pickard, 1983; Wolanski, 1986d).

Similar processes apparently control the temperature fluctuations on the shelf of the far southern Great Barrier Reef between 23 and 24°S (Griffin et al., 1987).

On the mid-shelf, no significant variations were reported at diurnal and tidal frequencies. At the shelf break, temperature fluctuations at tidal frequency reached 1.5°C. Large fluctuations of up to 4°C at diurnal frequency in shallow near shore sites were observed. The fluctuations in shallow water may be due to solar heating, and at the shelf break to tidal pumping. Low-frequency temperature fluctuations of up to 2.5°C for a 6- to 10-d period were spatially incoherent from 23 to 24°S. The seasonal fluctuations were typically 6°C.

Low-frequency currents, as opposed to tidally reversing currents, may also upwell cold water onto the shelf in the presence of reef passages if the sea level in the shelf is occasionally lower (by ≃ 0.1 m) than the adjoining Coral Sea. This sea-level difference could lift cold deep water up onto the passage and into the shelf by a "geostrophic pumping" process (Nof and Middleton, 1989). Presently available long-term current data (Thomson and Wolanski, 1984; Wolanski, 1986c; Wolanski et al., 1988a) support the internal tides and the tidal pumping mechanisms only.

3.6 TURBIDITY

Water turbidity is also found to be very patchy. As a result of wind waves resuspending the bottom sediment, coastal waters (depth <15 m) are turbid, with secchi disk visibility often less than 1 m (Wolanski et al., 1981). However, the spatial distribution of turbidity on the shelf is very variable, with patches of turbid water also found in offshore waters (Figure 3.15).

Patchiness is enhanced by the complex topography of coral reefs. Strong currents generate eddies in the lee of headlands, coral reefs, and islands. Eddy water is usually turbid, and this makes the eddies readily visible from the air (Figure 3.16). When the tidal currents reverse, these eddies are advected away by the prevailing currents and form a turbid water patch imbedded in clearer water (Figure 3.17).

FIGURE 3.15. Raw Landsat band 4 view of the central Great Barrier Reef on May 27, 1979, emphasizing water turbidity.

River plumes are generally very turbid, but this turbidity is also patchy. In the case of the Burdekin River plume during a flood, the concentration of fine particles (size <25 μm) within 60 km of the mouth of the river follows that predicted from dilution as derived from the salinity distribution. This finding suggests that the fine sediment may travel long distances (Wolanski et al., 1981). This does not apply to the coarse particles (size >25 μm) which settle near the mouth.

There is a similarity between the salinity distribution during Burdekin River floods and the distribution of terrigenous mud on the sea floor. This finding suggests that, once deposited on the mid-shelf, the riverine mud is fairly stable, probably because tidal currents are small and the Great Barrier Reef shelters the shelf from the oceanic swell. This sediment is resuspended in tropical cyclones. When tropical cyclone Winifred crossed the coast near Innisfail (17.5°S) on February 1, 1986, current metres moored on the inner shelf 80 km south of the cyclone track recorded currents peaking at 1.1 m s^{-1}, strong enough to erode the bed (Wolanski and Ridd, 1990). Gagan et al. (1987, 1988) analyzed the sediment on the sea floor before and after the passage of the cyclone. They showed that the passage of the cyclone produced a mixed terrigenous-carbonate storm bed 10 to 15 cm thick and extending to the 40-m depth contour located 30 km offshore. This storm layer resulted from *in situ* resuspension of the shelf sediment accompanied by shoreward transport of mud. One year later the storm layer was still preserved at a water depth <20 m, but was completely bioturbated further offshore. The bulk of the terrestrial plant detritus from the Johnstone River, a small river which flooded following the cyclone, settled within 2 km, and none of this detritus was found more than 15 km offshore. The authors suggested that terrestrial sediment would only reach the Great Barrier Reef by two means: in the plume of large Burdekin River floods, and following resuspension and advection during tropical

FIGURE 3.16. Aerial photograph of an eddy shed by a headland in the Whitsunday Islands area. The arrow points to the foam line or slick.

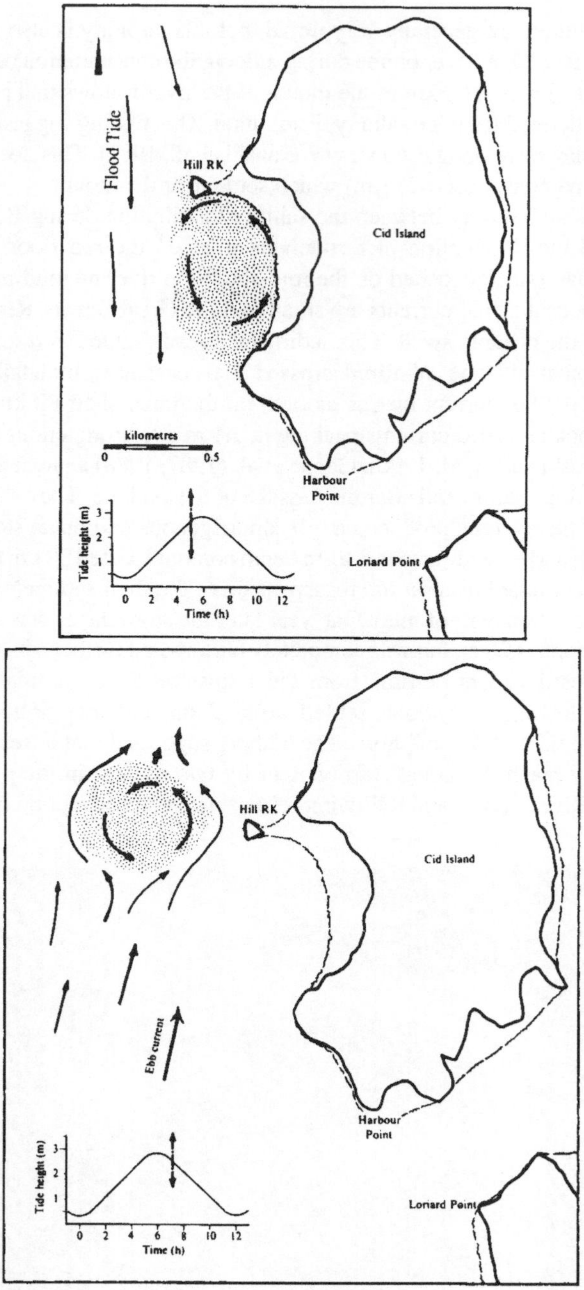

FIGURE 3.17. Sketch of the ejection of an eddy past a headland at reversing tidal currents. (From Hamner and Hauri, 1977. Reproduced with the kind permission of CSIRO Editorial Services.)

cyclones. Terrestrial sediment reaches the Great Barrier Reef preferentially in areas where the topography generates strong cross-shelf currents (Davies and Hughes, 1983; Sahl and Marsden, 1987).

Geomorphological studies miss out events of sediment initially settling on the reefs but then advected away later in a storm. Such events provide a transient stress to coral reefs, about which little is known. For instance, field studies by G. Jones (personal communication) in Cleveland Bay, Townsville, showed that a thin layer of suspended sediment settled on the fringing coral reefs of Magnetic Island, off Townsville, following dredging, but that the bulk of that sediment was later reexported in storms.

Only in the case of the Fly River estuary in the Gulf of Papua have the transport pathways of suspended sediment been studied from the river mouth to offshore. In that case it was found that the cohesive sediment is only exported offshore in large quantities during storm events, and that the sediment is carried in the river plume as small flocs. There have been no studies of flocculation dynamics in the Great Barrier Reef itself.

Wave-driven sediment resuspension is believed to be an important source of nutrient enrichment in the coastal waters of the Great Barrier Reef (Walker and O'Donnell, 1981). Following nutrient enrichment from cyclone Winifred, a diatom-dominated phytoplankton bloom developed within a few days in shelf waters, while dinoflagellate and flagellate populations did not bloom, a finding implying that mechanisms other than nutrient enrichment are necessary for red tide blooms (Furnas, 1989).

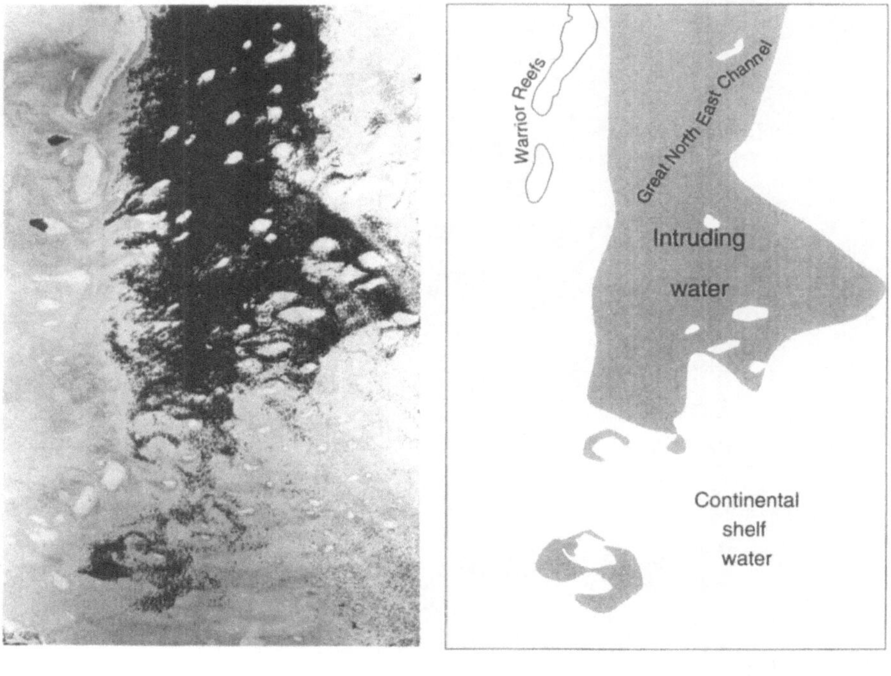

FIGURE 3.18. Landsat band 4 image of the Torres Strait, enhanced to eliminate bathymetric effects and to highlight differences in water color.

In the far-northern Torres Strait, three distinct water masses mix. All are turbid but with a different color (Figure 3.18). They are, respectively, the intruding, brackish, Gulf of Papua waters, the receiving Torres Strait water east of the Warrior Reefs, and the Torres Strait water west of the Warrior Reef. The latter are particularly turbid because of tidally driven sediment resuspension (Mulhearn, 1989).

3.7 BIOLOGICAL PATCHINESS

The distribution of chlorophyll is also very patchy at all scales, particularly near coral reefs (Figure 3.19). Temporal variability was demonstrated by pronounced differences in mean chlorophyll concentrations between transects. Clearly, nonautomated techniques using discrete sampling leads to biased results. At even smaller spatial scales around the coral reefs, the distribution of plankton, eggs, and larvae is also very patchy, the bulk of the material being concentrated in slick lines (Wolanski and Hamner, 1988).

Patchiness in the form of streakiness is commonly found in the distribution of floating organisms, such as the blue-green algae *Trichodesmium* (Furnas, 1992) and coral eggs (Figure 3.20; Oliver and Willis, 1987; Wolanski and Hamner,

FIGURE 3.19. Cruise tracks and near-surface temperature and chlorophyll concentration in the Ribbon Reefs area. (From Liston et al., 1992. Reproduced with the kind permission of Pergamon Press.)

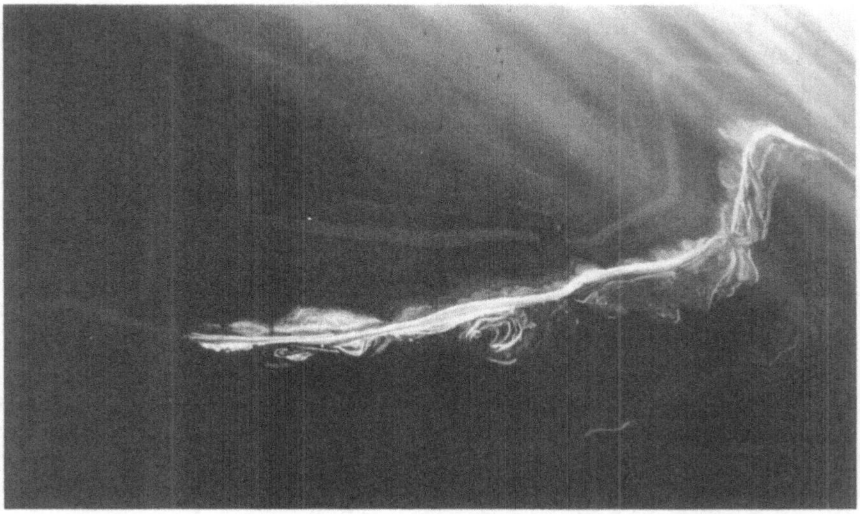

FIGURE 3.20. A few hours after mass spawning, coral eggs are aggregated in a coral slick which can be only 1 to 2 m wide but several hundreds of metres long.

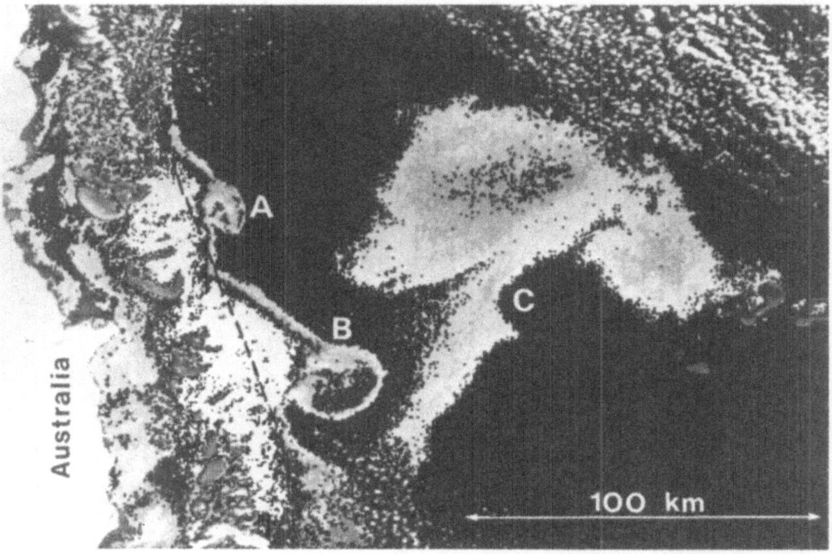

FIGURE 3.21. Coastal Zone Color Scanner view of the central region in the chlorophyll band. The dashed line is the 100-m depth contour. A to C are discussed in the text.

1988). These organisms are usually found aggregated in slicks that can be only a few metres wide but hundreds of metres long for coral eggs and several kilometres long for *Trichodesmium*. Patchiness in the distribution of plankton and fish larvae has long been recognized as a major unresolved problem in estimating populations (e.g., Kingsford et al., 1991; Victor, 1984; Williams and English, 1992). For nonmobile organisms, patchiness at scales larger than a few metres is probably due to oceanographic processes.

Satellite pictures (e.g., Figure 3.21) of the Great Barrier Reef in the chlorophyll band show two bands of high chlorophyll concentration, one near the coast and one within the Great Barrier Reef (Wolanski, 1986d; Gabric et al., Nof and Middleton, 1989). The chlorophyll distribution changes markedly between successive satellite passes, demonstrating that the chlorophyll distribution is highly variable. The chlorophyll-rich coastal band may be due to nutrient enrichment from the sediment stirred by the wind (Walker and O'Donnell, 1981). The chlorophyll-rich strip in the Great Barrier Reef matrix may be due to a number of phenomena, such as tidal pumping and leaching of nutrients and detritus from coral reefs. What is striking in these images is the high degree of patchiness in the chlorophyll distribution in the Great Barrier Reef matrix at all scales, from tens of kilometres down to the individual pixel of the picture. At times the chlorophyll patchiness appears to be due to readily identifiable oceanographic phenomena, such as in Figure 3.21 the two eddies (A and B) located on the continental slope and a mushroom jet (C) in the Coral Sea.

Furnas and Mitchell (1986, 1987) suggested that the phytoplankton population and productivity on the outer shelf of the Great Barrier Reef are very patchy and variable and showed no obvious association with shelf-scale upwelling events, although they may be stimulated by them (Andrews and Furnas, 1986).

4 The Tides

The tides were among the first processes to be studied. Tidal data are relatively easy to obtain with tide gauges and submerged water level recorders.

4.1 SEA LEVELS

The diurnal and semidiurnal tides are the dominant phenomenon controlling the sea level variations in the Great Barrier Reef. This is clear from a cursory examination of the spectrum of sea level (Figure 4.1), the dominant features of this figure being the peaks at diurnal and semidiurnal frequencies. There is also energy at periods >5 d; these are the low-frequency motions discussed in Chapter 5. The classical way to deal with tidal data is to extract the tidal harmonic constituents. Easton (1970), Pickard et al. (1977), Hamon (1984), and Andrews and Bode (1988) have carried out such analysis on various data sets and found that in all areas of the Great Barrier Reef, the dominant constituents are the O_1, P_1, K_1, N_2, M_2, S_2, and K_2 constituents. Until recently the data were mainly from coastal sites and the spatial coverage was poor. Recently the A.I.M.S. initiated a long-term data collection on the tides over the whole length of the Great Barrier Reef. The variation with latitude of the dominant M_2 and K_1 tidal constants, for both coastal and outer reef stations, was measured and is shown in Figure 4.2. This figure also shows the variation with latitude of the tidal range (the sum of the amplitudes of all the dominant tides) and of the tidal form factor, F (Foreman, 1977). F is used to quantify the relative amplitude of diurnal and semidiurnal tides; for $F < 0.25$ the tides are semidiurnal, for $F > 3$ the tides are diurnal, and for $0.25 < F < 3$ the tides are mixed. The tides of the Great Barrier Reef are mixed everywhere except near the coast in the southern region near Broad Sound (22.3°S), where they are semidiurnal. The tides show marked latitudinal and cross-shelf changes. Near Broad Sound there is a very large amplification of the semidiurnal tide, while the diurnal tide is much less amplified. Near Broad Sound the M_2 tide is the dominant constituent (average amplitude $a =$ 0.75 m), about twice that of the highest other constituents (about 0.35 to 0.4 m for S_2 and K_1; Hamon, 1984). Matthew Flinders, over 150 years ago, proposed that the amplification is due to the reefs located offshore and essentially blocking the inshore-offshore movement of the tides. This forces two tidal waves to propagate into the system toward Broad Sound, one from the south through Capricorn Channel and one from the north (offshore Townsville), where the reef density is least and where water can also readily enter the shelf.

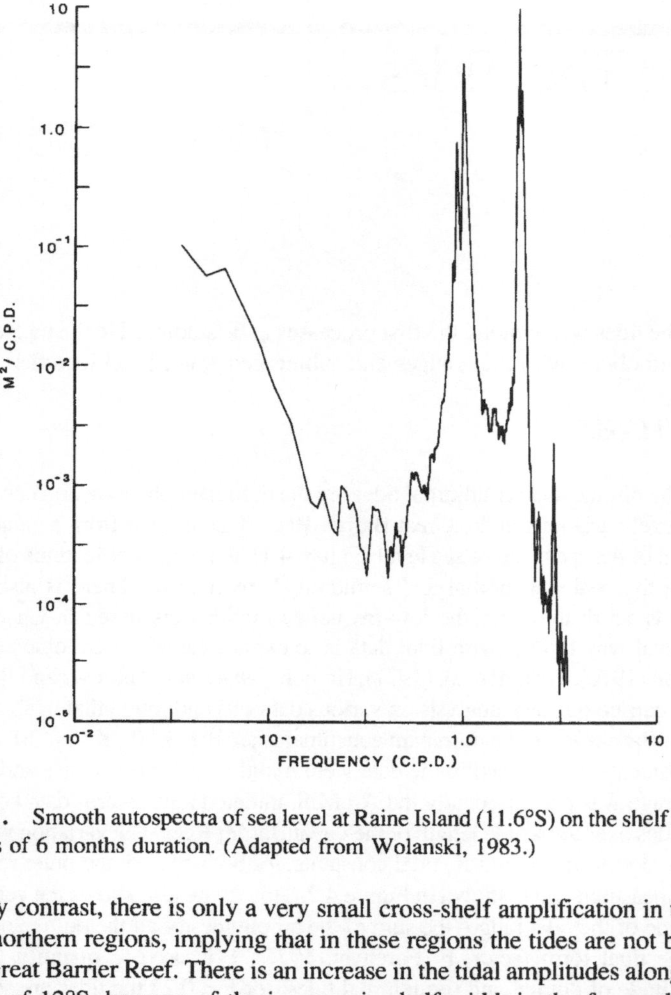

FIGURE 4.1. Smooth autospectra of sea level at Raine Island (11.6°S) on the shelf slope, from a time series of 6 months duration. (Adapted from Wolanski, 1983.)

By contrast, there is only a very small cross-shelf amplification in the central and northern regions, implying that in these regions the tides are not blocked by the Great Barrier Reef. There is an increase in the tidal amplitudes along the coast north of 12°S, because of the increase in shelf width in the far northern Great Barrier Reef (Wolanski, 1983). The situation is particularly complex in Torres Strait, which is influenced by the tides from both the Arafura Sea and the Coral Sea (Wolanski et al., 1988b). However, the Torres Strait prevents the tides from the Arafura Sea from entering more than 50 km into the Great Barrier Reef (Figure 4.3).

4.2 TIDAL CURRENTS

Obtaining tidal current data is much more difficult than obtaining tidal height data. Moorings are necessary, and are at risk from ships and fishermen. The complex topography makes it difficult to select a suitable site. Indeed, because of the channelization of the currents, two current metres a few kilometres apart on

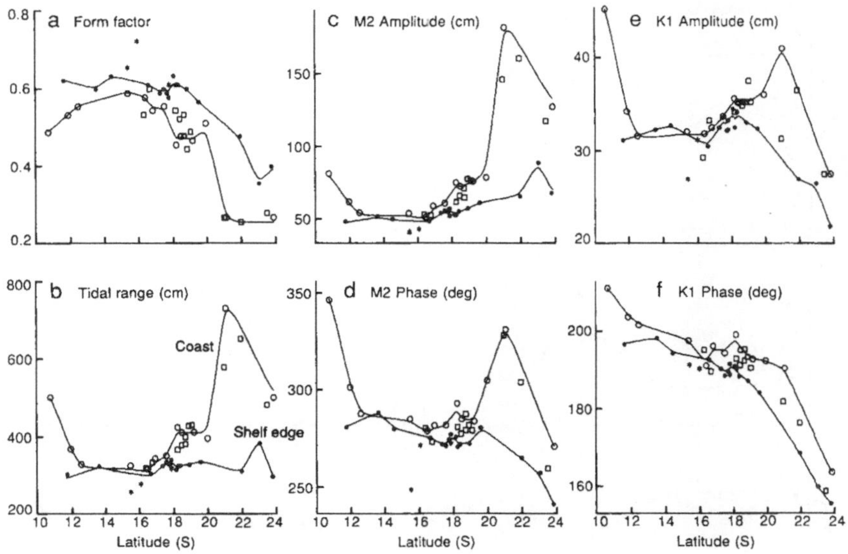

FIGURE 4.2. Variation with latitude of (a) the form factor, and the amplitude and phase of the (b) M_2 and (c) K_1 tidal constituents at the shelf break (·) and at the coast (O). (Andrews, personal communication.)

the shelf, one near a reef and one in interreefal waters, may measure very different tidal currents. In areas of high reef density, it is even open to question if unbiased or representative tidal current data can be obtained.

4.2.1 Central Region

Only in areas where the shelf is free of reefs can reliable tidal current data be obtained. This is one reason why the central region, and in particular Palm Passage, which has the least reef density, has been the focus of much research. Tidal ellipses are shown in Figure 4.4. The tidal currents are small in this region, the amplitude of the dominant M_2, S_2, K_1, and O_1 tides being only 0.12, 0.06,. 0.03, and 0.02 m s^{-1}, respectively. The tidal currents are primarily oriented cross-shelf on the outer shelf. Off Townsville (19.2°S), the currents lag the sea level by 90 to 100° in the manner of a standing wave in the middle of the study area. South of Townsville, the tidal amplitude increases as the tides become directed toward Broad Sound and behaves as a progressive wave.

The linear, analytical tidal models of Battisti and Clarke (1982a, b) for reef-free waters are able to reproduce the standing wave behavior of the tides off Townsville (Church et al., 1985). To explain the progressive wave behavior of the tides south of Townsville, Andrews and Bode (1988) used a depth-averaged, two-dimensional, linear, numerical model of the continental shelf. While such models qualitatively explain the large-scale behavior of the tides, the tidal circulation

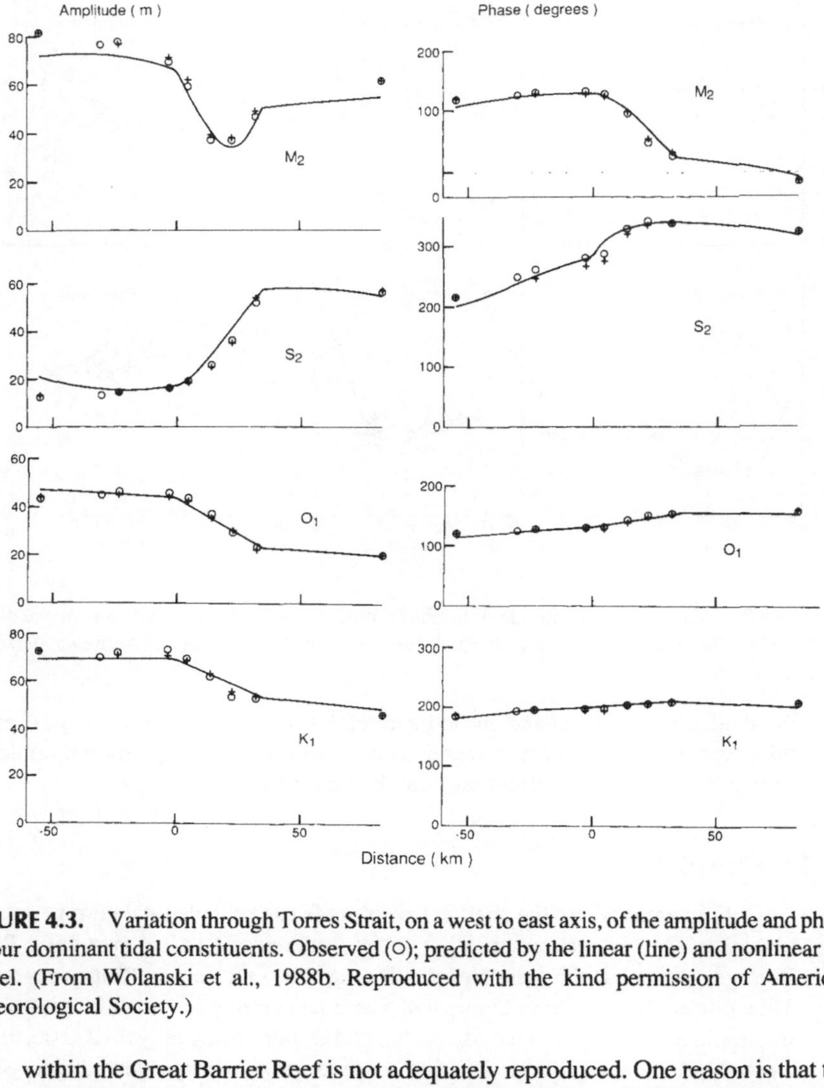

FIGURE 4.3. Variation through Torres Strait, on a west to east axis, of the amplitude and phase of four dominant tidal constituents. Observed (○); predicted by the linear (line) and nonlinear (+) model. (From Wolanski et al., 1988b. Reproduced with the kind permission of American Meteorological Society.)

within the Great Barrier Reef is not adequately reproduced. One reason is that the mesh size (5 nm $\simeq$ 9 km) is too coarse. Indeed, Figure 4.5 shows time series observations of currents at three sites a few kilometres apart in interreefal waters around Bowden Reef (19°S). The net longshore current was a northward with a magnitude of about 0.15 m s^{-1} at site 6. The tidal currents varied markedly from site to site, being nearly twice as large at site 11 than at site 5. In addition, there was a phase lag of up to 3 h between the tidal currents at these three sites, much larger than the phase lags of 1 min predicted by the model of Andrews and Bode (1988). Similar large spatial variations of the tidal currents over a few kilometres in interreefal waters were reported for other sampling periods by Wolanski and King (1990).

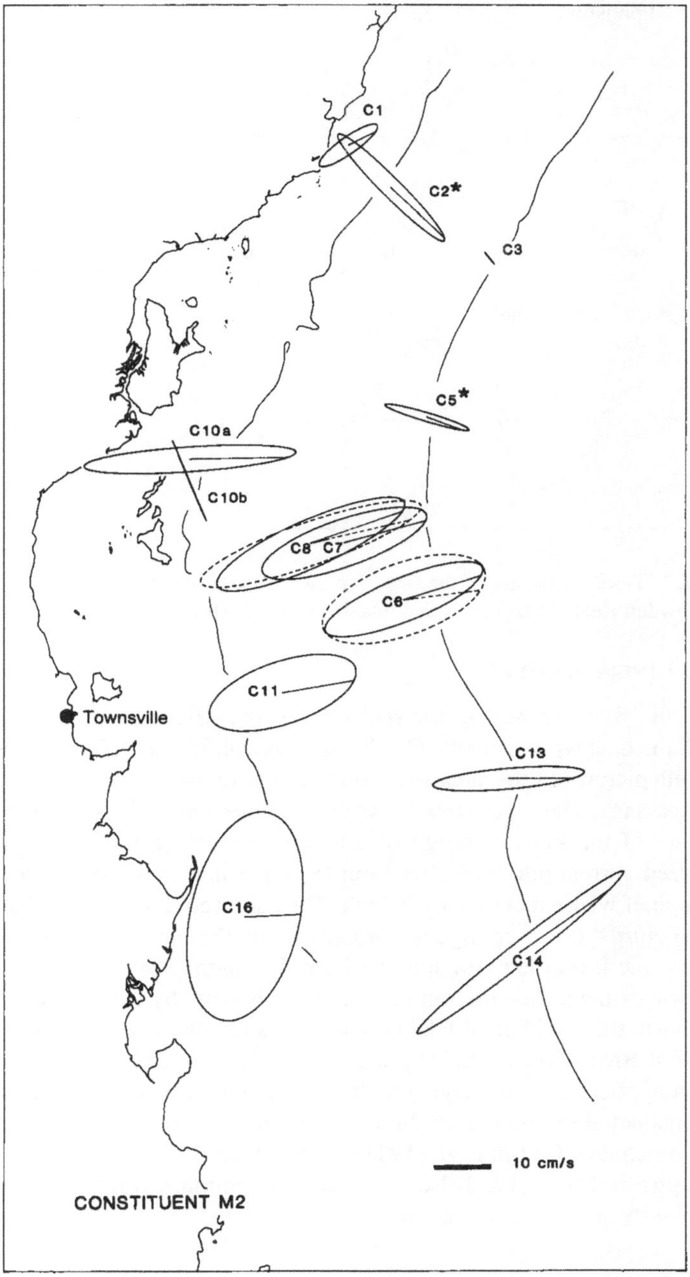

FIGURE 4.4. Observed current ellipses of the M_2 tide. (Adapted from Andrews and Bode, 1988.)

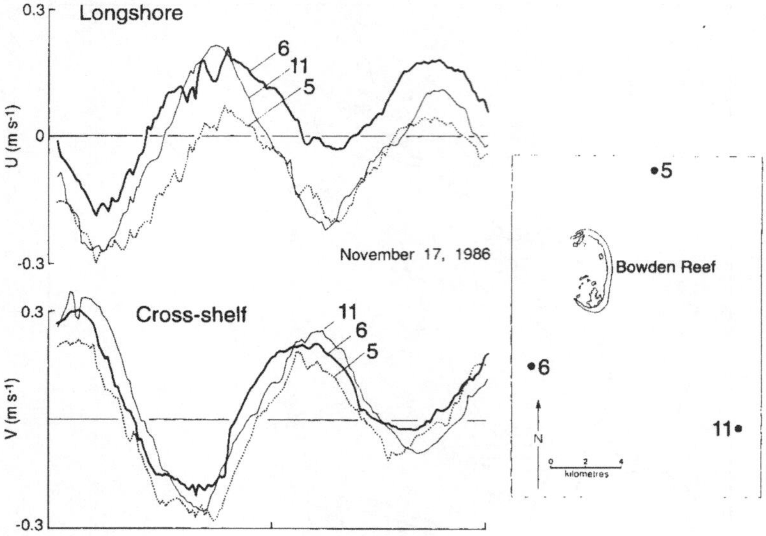

FIGURE 4.5. Time series plot of the observed currents on November 11, 1986, at mooring sites near Bowden Reef. (Adapted from Wolanski et al., 1989.)

4.2.2 Southern Region

In the southern region, the reef density is sufficiently high to effectively block the tidal wave from the Coral Sea (Figure 4.2). Color Zone Color Scanner satellite pictures in this area show tidal jet-vortex pair systems, one in front of each passage. These complex flows can be expected to dissipate a measurable fraction of the kinetic energy of the tidal currents. Middleton et al. (1984) analyzed current and tidal data from the inner and mid-shelf, away from the outer shelf where reef density is high. They showed that the semidiurnal flood tide (Figure 4.6 for the M_2 tide) propagates on the inner and mid-shelf as two waves, one from the north and one from the south. The convergence of these two waves magnifies the semidiurnal tidal heights by a factor of about four relative to the semidiurnal tidal height at the shelf break. In addition, the funnel shape of Broad Sound (22.3°S) and shallow waters produce a further amplification of about 1.5, convergent on the Broad Sound region. This amplification is frequency dependent; semidiurnal tides amplify most and diurnal tides amplify much less. Griffin et al. (1987) analyzed current-metre data south of 23°S in Capricorn Channel and also found that the semidiurnal flood tides propagate northwestward into the Lagoon.

4.2.3 Northern Region

Reefs and shoals are scattered, often densely, all over the shelf in this area. This can be expected to increase the resistance to flow. It also makes difficult the collection of tidal current data that are not affected by the presence of nearby reefs.

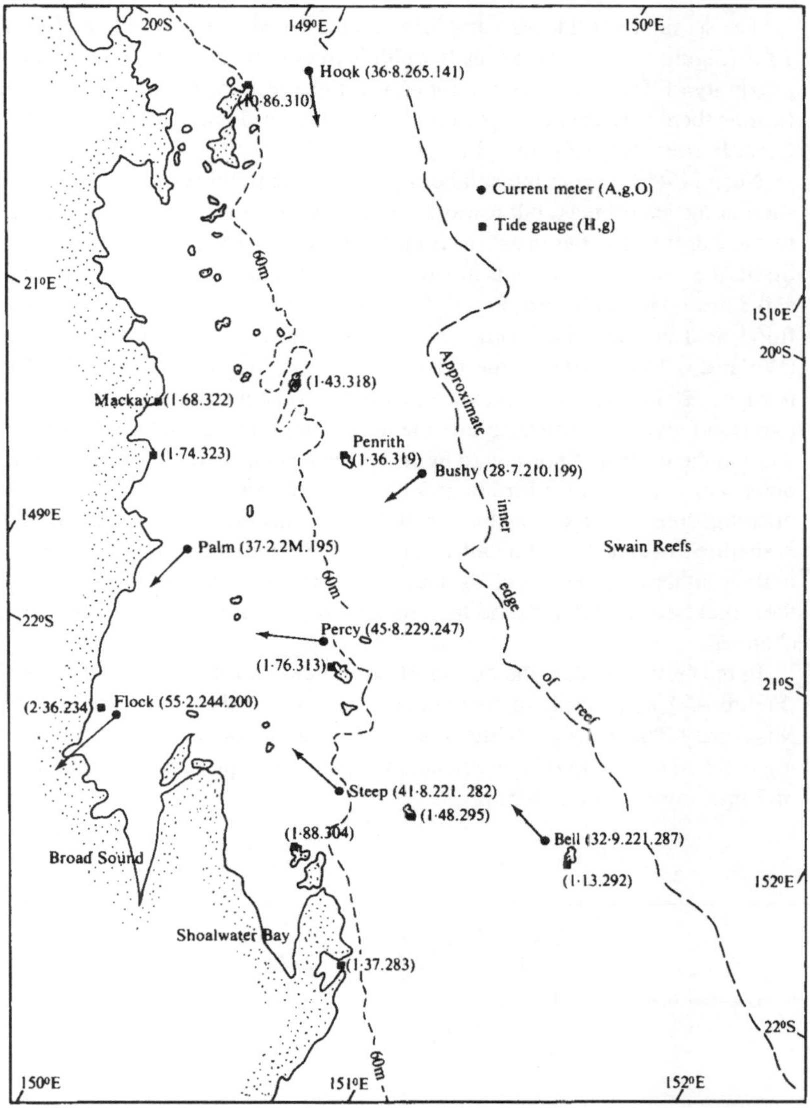

FIGURE 4.6. Observed flood tides and sea level amplitudes and phases in the southern region. The numbers in brackets refer, for sea level, to the amplitude (in metres) and the phase (GMT-10 h), and, for currents, to the tidal ellipse semimajor axis (cm s^{-1}), the phase, and the orientation (°) of the semimajor axis. (From Middleton et al., 1984. Reproduced with the kind permission of Pergamon Press.)

The available data (Figure 4.2) show that there is little cross-shelf amplification of the semidiurnal tides south of 12°S, except for a small amplification at Princess Charlotte Bay (14.3°S) which may be attributed to topographic steering by the funnel shape and shallow waters of Princess Charlotte Bay.

The semidiurnal tides are amplified on the Great Barrier Reef shelf north of 12°S (Figure 4.2), where the shelf width increases with decreasing latitude. The proximity of Torres Strait does not appear to be responsible for the tidal amplification there because the tides from the Arafura Sea do not propagate far into the Great Barrier Reef (Figure 4.3).

Near 11.4°S, i.e., in the southern part, the tidal ellipses are oriented cross-shelf at the shelf break, but toward the northwest near the coast, the tides flood to the north. Near the coast, at Shortland Reef (10.8°S) south of the Torres Strait, the flood tidal currents are also northward, the M_2 tidal currents peaking at 0.3 m s^{-1}. However, near Pearce Cay (9.5°S) in the Great North East Channel, flood tidal currents are southward, the M_2 tidal current peaking at 0.2 m s^{-1} (Wolanski, 1983). Hence, the tidal behavior on the Great Barrier Reef shelf north of 12°S duplicates somewhat that in the southern Great Barrier Reef, with two flood waves propagating into the area. The semidiurnal tide floods, southward in the north and northward in the south, meet near 10.5°S. In this area, the outer reef is continuous for about 90% of the distance at the shelf break, with openings (reef passages) over about 10% of the distance. Further, the shelf itself is shallow, with depths of about 10 to 30 m with numerous shoals and reefs. As in the southern region, the cross-shelf propagation of the semidiurnal tide from the Coral Sea is hindered, and the tide instead is channelized through reef-free channels.

In the Gulf of Papua, the dominant semidiurnal tides are oriented cross-shelf (Figure 4.7) as a result of topographic steering by the Fly River estuary. In Missionary Passage (site E), the tidal currents are topographically steered, peaking at 1.5 m s^{-1}, the flood tide oriented westward, i.e., propagating from the Gulf of Papua toward Torres Strait.

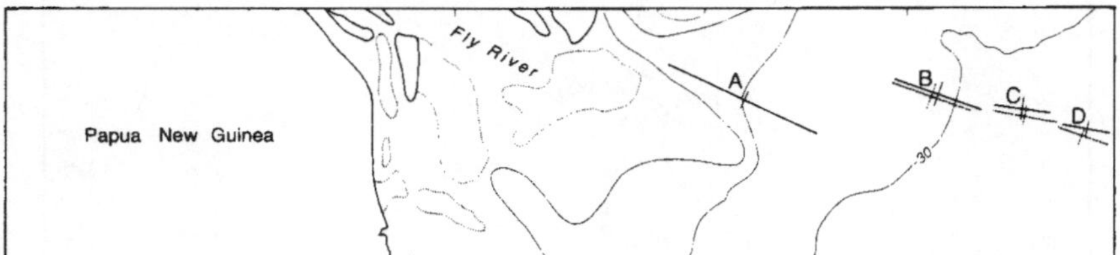

FIGURE 4.7. M_2 tidal ellipses in the Gulf of Papua near the mouth of the Fly River. (Adapted from Wolanski et al., 1992c.)

4.2.4 Tidal Blocking by Reefs: An Example in Torres Strait

The reefs of Torres Strait (Figure 1.2d) form a narrow wave guide, separating the Gulf of Carpentaria to the west from the Great Barrier Reef continental shelf to the east. Gradients of the tidal variation through Torres Strait have been measured in great detail in that area, much more intensively than in any other area of the Great Barrier Reef, and these data can be used to gain insight in the blocking of the tidal propagation by coral reefs. Slightly to the east of Torres Strait, e.g., at sites E and G, the tidal currents are oriented east-west (Wolanski et al., 1988b). The problem of tidal propagation in Torres Strait is thus essentially a one-dimensional phenomenon, the dominant axis of tidal variation being oriented east-west. Torres Strait can then be viewed as a narrow channel constricting the water exchange between the Arafura Sea and the Great Barrier Reef continental shelf (Figure 4.8). Incident tidal waves from both seas propagate toward Torres Strait. A fraction of these waves is reflected; the rest propagates into Torres Strait. This situation was modeled by Wolanski et al. (1988b), who used the continuity equation and the linearized momentum equations with a linear friction law for each of the three regions (i = 1, 2, 3) shown in Figure 4.8,

$$\frac{\partial u_i}{\partial t} + g\frac{\partial \eta_i}{\partial t} + K_i\frac{u_i}{H_i} = 0 \tag{4.1}$$

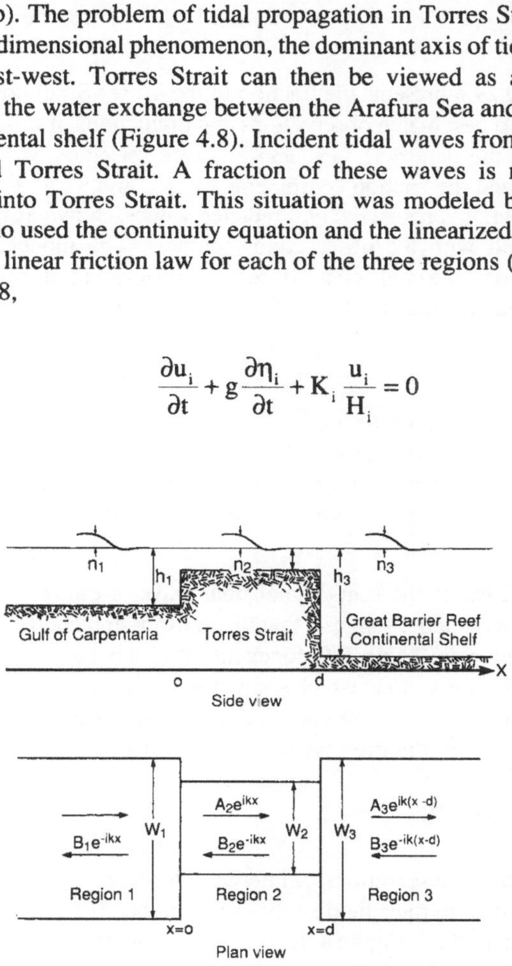

FIGURE 4.8. Sketch of the Torres Strait waveguide model geometry. (From Wolanski et al., 1988b.)

$$H_i \frac{\partial u_i}{\partial t} + \frac{\partial \eta_i}{\partial t} = 0 \quad (4.2)$$

where x is the distance, t the time, g the acceleration due to gravity, η the sea level and u the east-west velocity. The boundary conditions are the known tidal constants at the Arafura side of Torres Strait (x = 0) and at the Great Barrier Reef side of Torres Strait (x = d). Conservation of mass requires that the water fluxes have to be the same, i.e.

$$Fu_1 = u_2 \text{ at } x = 0, \text{ and } Fu_3 = u_2 \text{ at } x = d \quad (4.3)$$

where 1/F = 0.29 represents the fraction of the Torres Strait not blocked by reefs and islands. To best match observations and predictions of tidal variations across Torres Strait, the depths H_1, H_2, and H_3 were set equal to 11, 10, and 13 m, respectively, and the friction coefficients K_1, K_2, and K_3 to 2.5×10^{-3}, 6×10^{-3}, and 2.8×10^{-3} m s^{-1}, respectively. Predicted values, shown as full lines in Figure 4.3, agree well with the observations (circles). The model suggests that for semidiurnal tides, the coefficients of reflection (R) and transmission (T) for tidal waves propagating from the Gulf of Carpentaria to the Great Barrier Reef shelf are:

$$\begin{aligned} R &= 0.68 \times e^{i6°} && \text{for semidiurnal tides} \\ &= 0.63 \times e^{-i11°} && \text{for diurnal tides} \\ T &= 0.29 \times e^{i45°} && \text{for semidiurnal tides} \\ &= 0.36 \times e^{i36°} && \text{for diurnal tides.} \end{aligned} \quad (4.4)$$

Hence, only 29% of the semidiurnal tidal wave is transmitted through Torres Strait. This linear model suggests that the transmission coefficient increases with decreasing frequency, being 71% for events of 16 d duration. This prediction does not agree with observations which show that low-frequency events are unable to propagate across Torres Strait (see Chapter 5). The inability of the linear model to cope with the long-duration events is shown later to be due to nonlinear effects.

4.3 INTERNAL TIDES

The first direct observations of internal tides were reported by Wolanski and Pickard (1983) from their field study in September 1980 at the shelf break near Myrmidon Reef (18.3°, Figure 4.9). The well-mixed surface layer in the Coral Sea was 120 m thick, with the thermocline located below. A string of six current metres, each with a temperature sensor, was moored in September 1991 near site E for 50 d at the shelf break (110-m depth). The metres were located between 27

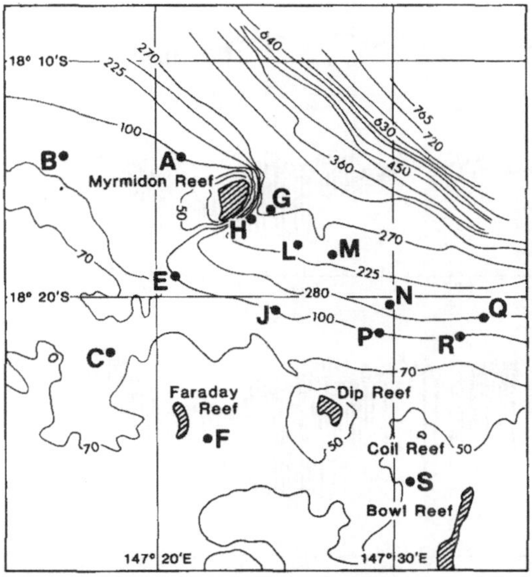

FIGURE 4.9. Bathymetry (in metres) around Myrmidon Reef and mooring sites.

and 104 m below the surface, i.e., the bottom sensor was 6 m above the bed. The current metre data showed large temperature fluctuations at the semidiurnal frequency (Figure 4.10). At the 27-m depth, the temperature was nearly constant. At the 104-m depth, there were large (up to 8°C) temperature fluctuations. These occurred at tidal frequency, although the tides and the temperature signal showed no unique phase relationship. The tidal currents were oriented about cross-shelf, flooding toward the coast. The tidal currents were largest at the 27-m depth and peaked at about 0.5 m s^{-1} for the semidiurnal component. A comparison of the temperature time series at the 104-m depth with the temperature profiles implies that the vertical excursions of the thermocline were as much as 110 m, the thermocline normally located at the 120-m depth moving up to between 27 and 55 m of the surface.

These internal tides are among the largest reported worldwide. They may be very important ecologically, as they generate a daily upwelling. To understand their dynamics, a detailed field study was carried out by Wolanski (1986d), who moored current metres and temperature sensors at 14 sites around Myrmidon Reef (Figure 4.9), both in the well-mixed layer and in the stratified layer below. In addition, cross-slope transects of temperature were measured. These transects (Figure 4.11) show that the internal tides slosh up and down the continental slope, trapped along the slope, as an internal wave at semidiurnal frequency. The vertical displacements of the thermocline are largest at the shelf break and diminish further offshore. Maximum vertical displacements were about 100 m just in the lee (for the East Australian

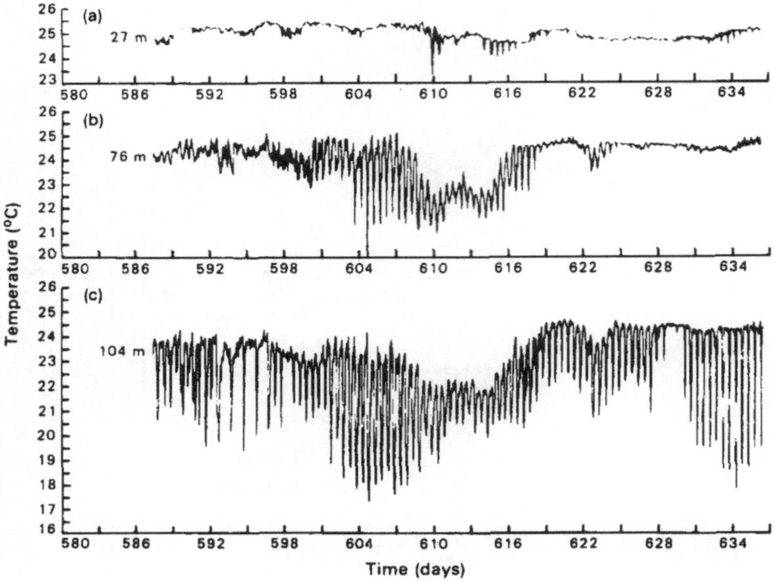

FIGURE 4.10. Time series plot of temperature at site E near Myrmidon Reef at three depths for August to September 1981. (From Wolanski and Pickard, 1983. Reproduced with the kind permission of CSIRO Editorial Services.)

Current) of Myrmidon Reef, but were smaller further upstream or downstream of Myrmidon Reef (Figure 4.12a), mimicking the behavior of the M_2 tidal current, which was largest in the lee of Myrmidon Reef because of topographic steering (Figure 4.12b). This finding suggests that the exceptionally large internal tides near Myrmidon Reef are due to a combination of purely internal tides sloshing up and down the continental slope, and to topographic effects in the lee of Myrmidon Reef in the presence of the East Australian Current. The internal tide is largely a mode 1 internal tide following the definition of Baines (1982), i.e., it raises and lowers at tidal frequency the thermocline separating the well-mixed upper layer from the deeper waters. The internal tide ray's slope is nearly identical to the shelf slope between 100 and 300 m. At large depth, the shelf slope is much larger. This implies that the internal tide generated at a depth >300 m cannot propagate toward the shelf break and is reflected offshore. Thus, only the continental slope at a depth <300 m contributes to the internal tide at the shelf break. There were also mode 2 internal tides, i.e., motions within the stratified waters below the thermocline without a movement of the thermocline.

Mode 1 internal tides motions were calculated from the data assuming a two-layer ocean as sketched in Figure 4.13. The barotropic (–) and baroclinic (′) components of the horizontal cross-shelf currents u_i(i = 1, 2 in the top and bottom layers, respectively, were calculated from the current metre data using the relations

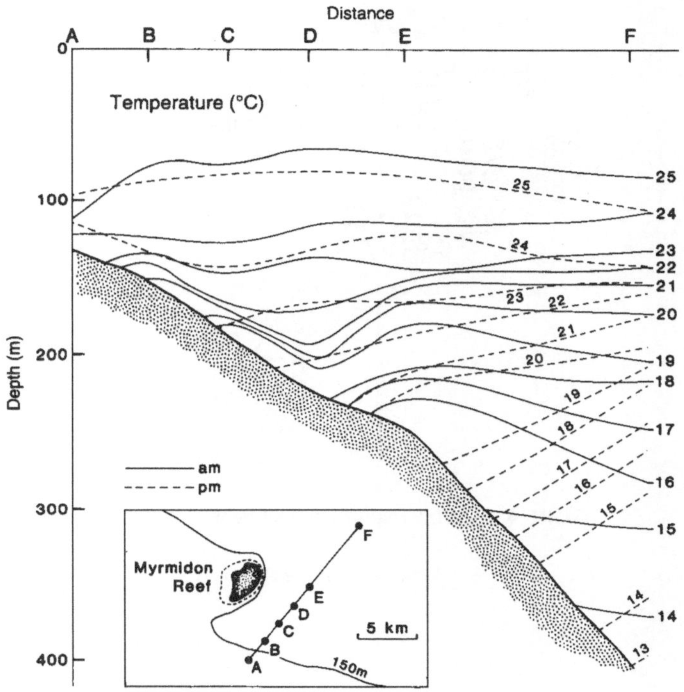

FIGURE 4.11. Cross-slope distribution of the temperature on a cross-slope transect near Myrmidon Reef. (Adapted from Wolanski, 1986d.)

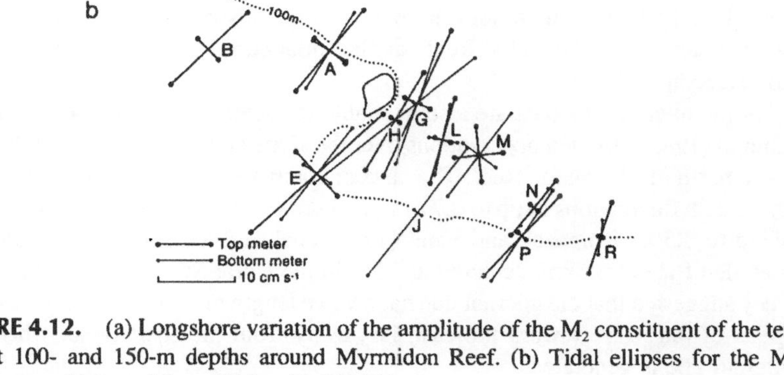

FIGURE 4.12. (a) Longshore variation of the amplitude of the M_2 constituent of the temperature at 100- and 150-m depths around Myrmidon Reef. (b) Tidal ellipses for the M_2 tidal constituents. (Adapted from Wolanski, 1986d.)

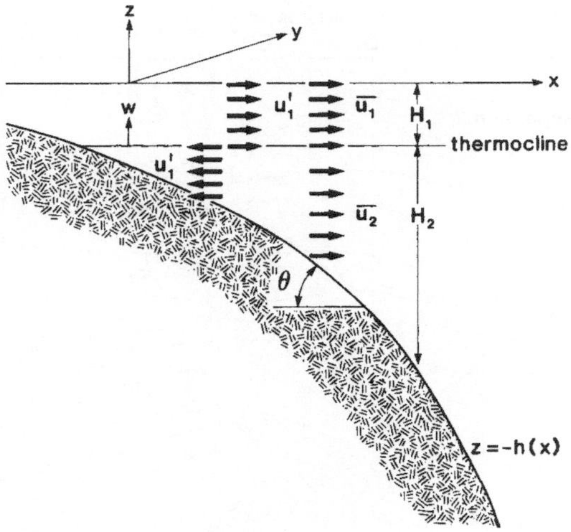

FIGURE 4.13. Internal tide model geometry.

$$H_1 u_1' + H_2 u_2' = 0 \tag{4.5}$$

$$\bar{u}_1 = \bar{u}_2 \tag{4.6}$$

The vertical barotropic current w at the elevation of the thermocline ($z = -H_1$) can also be calculated from the current metre data from the continuity equation

$$\bar{w} = -\bar{u}\tan\theta - (h + z)\,\partial\bar{u}/\partial x - (h + z)\,\partial\bar{v}/\partial y \tag{4.7}$$

where h is the total depth and θ is the slope of the sea floor. The vertical excursions of the thermocline on the continental slope, calculated from Equation 4.7, were 50 m due to the slope of the sea floor (the term tan θ), 35 m due to cross-shelf advection, and 15 m due to longshore advection. At site R at the shelf break, far downstream from Myrmidon Reef, smaller tidal currents result in internal tides not exceeding 30 m.

Some internal tide data are also available for deeper waters in the Coral Sea. Church (1983) reported observations over the slope at 270 m in a 300-m depth at a site north of Myrmidon Reef. The current components were also largely cross-shelf, with fluctuations of up to 0.7 m s^{-1} and semidiurnal temperature oscillations of up to 3.5°C. Tomczak and Fang (1983) analyzed temperature data along a triangle of about 55 km, centered at 18°S in the Coral Sea off Myrmidon Reef. They suggested that the internal tide has a wave length of about 28 km and a phase speed of 0.6 m s^{-1} toward the east, i.e., away from the shelf break where the internal tide is generated.

In the northern Great Barrier Reef, the slope of the sea floor on the shelf slope is much steeper than the slope of the internal wave rays, so that internal tides generated on the slope cannot propagate onto the shelf. Upwelling at tidal period in such areas is restricted to the vicinity of reef passages and is due to the presence of strong tidal currents through reef passages. Upwelling at tidal period is evidenced by the temperature fluctuations in reef passages (Figure 3.14). Thompson and Golding (1981) proposed an elegant and simple analytical model for this upwelling. The model predictions are very sensitive to details of the topography, and it is only recently (Wolanski et al., 1988b) that bathymetric and oceanographic data have been available for testing the model at the passage between Ribbon Reefs No. 3 and 4 (15.5°S; Figure 4.14). The sea floor slope offshore from the passage is an order of magnitude larger than the slope of the internal wave rays, so internal tides cannot propagate onto the shelf from the Coral Sea. The cross-section of the passage is funnel shaped and can be approximated by the solid lines in Figure 4.14. With the x-axis oriented as in this figure, with x = 0 being at the sill, the velocity u in a cross-section of the passage is, by continuity,

$$u(x) = u(x = 0)H(x = 0)W(x = 0)/H(x)W(x) \quad (4.8)$$

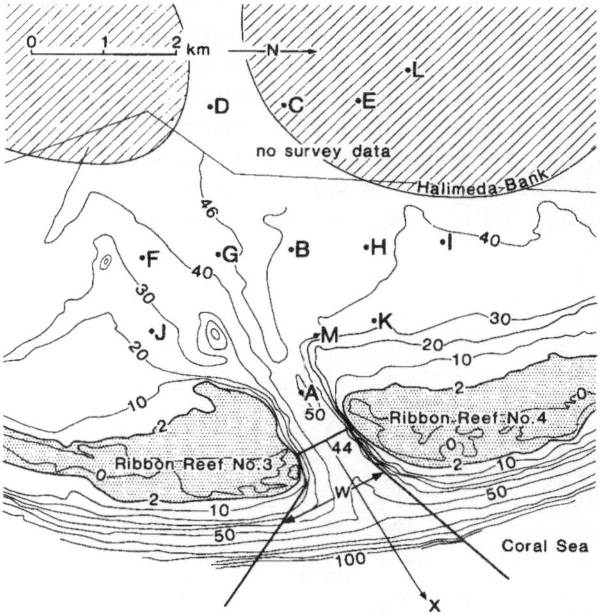

FIGURE 4.14. Bathymetry (in metres) around Ribbon Reefs No. 3 and 4, location of mooring sites, and modeled geometry of the reef passage. The dotted areas are the shallows (nearly emergent at spring low tide); the striped areas are the *Halimeda* meadows. (Adapted from Wolanski et al., 1988a.)

where H is the depth and W the width of the channel. Tidally induced upwelling occurs in the region of supercritical flow, so no wave-induced shear exists between the homogeneous upper layer and the temperature-stratified water below, as sketched in Figure 3.13b. The region of supercritical flow is shoreward of the point where u = c, where c is the fastest speed of internal gravity waves, determined by the dispersion relation (Thompson and Golding, 1981)

$$NH_1/c + \tan(Nh/c) = 0$$

$$\text{where } N = \sqrt{\frac{g}{\rho}\frac{d\rho}{dz}} \tag{4.9}$$

is the Brunt-Vaisala frequency in offshore waters below the well-mixed surface layer, g is the acceleration due to gravity, ρ is the density determined by temperature, z is the vertical axis, and H_1 and h are the thicknesses of the surface mixed layer and the stratified layer entrained upward, respectively. Estimates of c(x) from Equation 4.8 are shown in Figure 4.15. The velocity u(x = 0) was measured by a current metre deployed in the passage at site A. From that data, u(x) can be calculated from Equation 4.9, and the results are also plotted in Figure 4.15 for two different assumptions: a constant-width passage (W(x) = constant) and a funnel-shaped passage. The intersect of curves u(x) and c(x) gives the maximum depths from which water can be sucked, in this case, 88 m for the constant-width passage and 62 m for the funnel-shaped entrance. The predictions of upwelling of 62 m compared favorably with the observed ones of 65 to 68 m (Wolanski et al., 1988a), implying that the model of Thompson and Golding (1981) yields realistic results if the bathymetry is accurately included in the model.

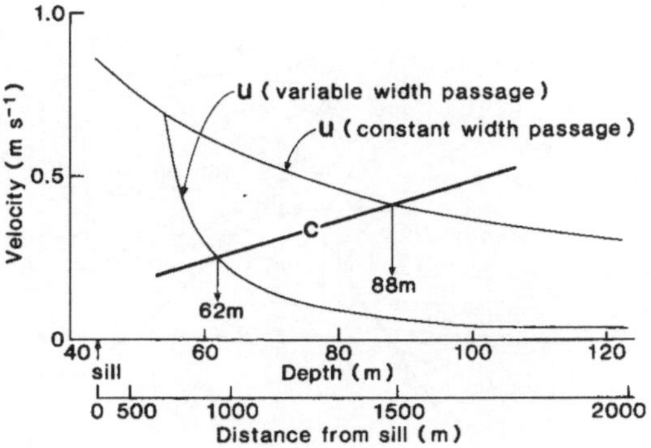

FIGURE 4.15. Plots of c(x) and u(x), where x is the distance offshore from the reef passage between Ribbon Reefs No. 3 and 4. (From Wolanski et al., 1988a. Reproduced with the kind permission of Academic Press.)

FIGURE 4.16. Aerial photograph of the tidal jet at Ribbon Reefs No. 3 and 4.

Shoreward of the passage, a mushroom jet (a tidal jet-vortex pair) develops at flood tide, as sketched in Figure 3.13a, and is readily visible (Figure 4.16).

Near-bottom temperature time series at the mooring sites on the shelf show that the upwelled water moves shoreward, as a near-bottom layer entrained in the tidal jet, and can readily be tracked as it moves from the Passage to at least 4.2 km shoreward.

Nutrients are also upwelled and are advected with the tidal jet not to the coral reefs on either side of the passage, but to the meadows of the calcareous green alga *Halimeda*. These meadows are located several kilometres shoreward from the reef passage (Figure 4.14). The quantity of nitrogen upwelled is sufficient to supply the total nitrogen requirements of the *Halimeda* meadows (Wolanski et al., 1988a).

The meadows of *Halimeda* on the shelf near the Ribbon Reefs may then be maintained by the tidal pumping mechanism. Large meadows of *Halimeda* have now been found to be common (E. Drew. personal communication) in other areas of the Great Barrier Reef where upwelling by tidal pumping or internal tides is known to be a common occurrence, such as south of Myrmidon Reef (18.3°S) and near Raine Island Entrance (11.6°S). This suggests that in these areas the tidally driven upwelling may become excessive for the maintenance of a healthy coral reef community. This finding follows that of Roberts et al. (1992), who found that excessive upwelling on the Nicaraguan shelf also induces a negative response in the coral-building process and favors instead algal communities.

5 Low-Frequency Motions

Low-frequency (LF) motions refer here to the motions which have a period greater than a few days, and which are not driven by the astronomical tides. They are calculated numerically from time series of wind, currents, and sea-level data by filtering out the tidal components.

5.1 LOW-FREQUENCY SEA-LEVEL OSCILLATIONS

Particularly during the southeast trade winds, when the wind is coherent over the whole length of the Great Barrier Reef, the LF fluctuations in sea level appear coherent with each other over long distances. This is illustrated in Figure 5.1, showing a time series of the LF sea-level fluctuations from Gladstone (23.9°S) to Carter Reef (14.6°S). The LF sea-level fluctuations had a maximum range of 0.35 m, about on tenth of the tidal range. Also plotted in Figure 5.1 are the LF atmospheric pressure and the longshore wind component at a number of sites. The LF longshore wind and sea level are apparently highly correlated, one with the other. No more than one fourth of the LF sea-level fluctuations can be accounted for by the atmospheric pressure.

The autospectra of LF sea-levels (Figure 2.4) show that the energy density generally increases with increasing period (decreasing frequency) for periods >4 d, i.e., there are no clear preferred frequencies. There is also a number of small, broad peaks between 10- and 200-d periods. Taking into account that the scales in this figure are logarithmic, the greatest variances (the largest amount of energy) occur at periods of a few days to a few weeks. This is the "weather band", i.e., the shelf water response to atmospheric fluctuations.

The LF fluctuations in sea level are largely trapped on the shelf. This is apparent from a cursory examination of a time series plot (Figure 5.2) of the LF sea level in the Coral Sea at Flinders Reef (17.7°S), where the fluctuations were about a few centimetres peak to trough, and on the Great Barrier Reef mid-shelf at Britomart Reef (18.3°S), where they were 30 cm.

In the central Great Barrier Reef, these LF sea-level fluctuations propagate northward at a speed of about 450 km/d^{-1}, as shown by the coherence and phase relationships between sea level at Townsville (19.2°S) and at Carter Reef (14.6°S; Figure 5.3a). The coherence is especially high at periods between 15 and 70 d, but significant at the 90% confidence limit for all periods >10 d.

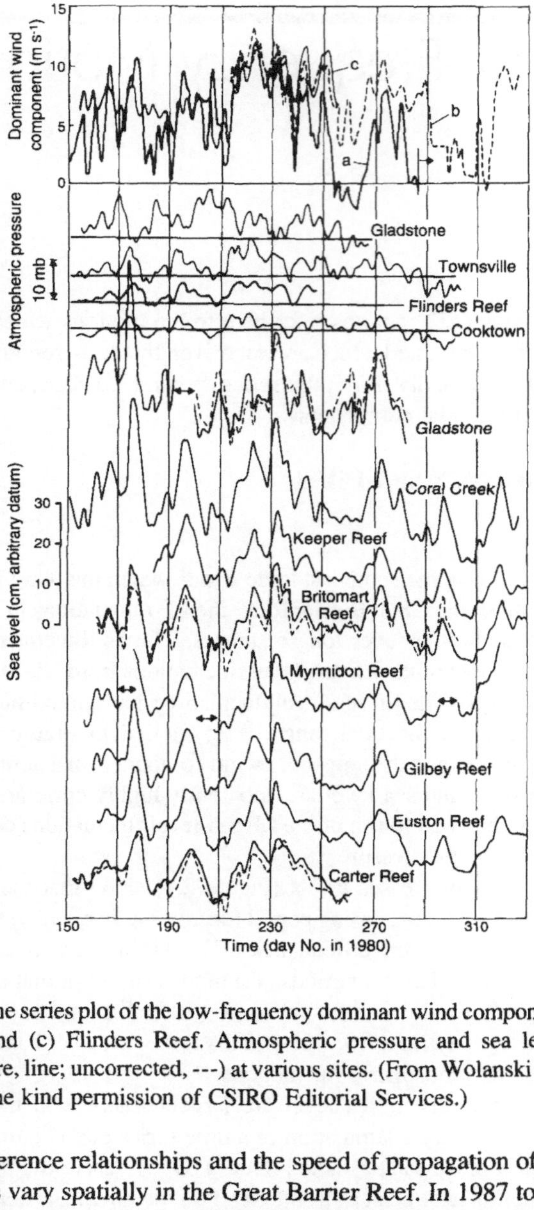

FIGURE 5.1. Time series plot of the low-frequency dominant wind component at (a) Rib Reef, (b) Carter Reef, and (c) Flinders Reef. Atmospheric pressure and sea level (corrected for atmospheric pressure, line; uncorrected, ---) at various sites. (From Wolanski and Bennett, 1983. Reproduced with the kind permission of CSIRO Editorial Services.)

The coherence relationships and the speed of propagation of the LF sea-level fluctuations vary spatially in the Great Barrier Reef. In 1987 to 1988, a network of tide gauges was maintained throughout the length of the Great Barrier Reef, with a maximum separation distance of about 1600 km between Gannet Cay (22°S) and Raine Island (11.6°S). These data were analyzed by Andrews (personal communication). The LF sea-level fluctuations were most energetic in the weather band of a 13- to 40-d period. He calculated the variation of the coherence squared c_o^2 as a marker, the data suggest that the disturbances propagate coherently only 400 km in the southern Great Barrier Reef, 800 km in the central Great

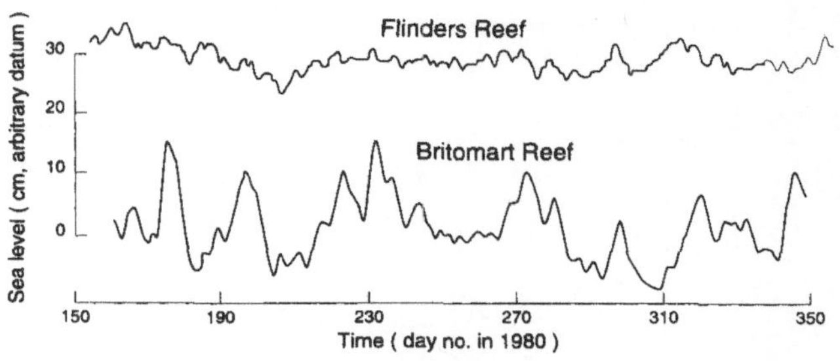

FIGURE 5.2. Time series plot of low-frequency sea level at Flinders Reef (in the Coral Sea) and at Britomart Reef (on the continental shelf). (From Wolanski and Bennett, 1983. Reproduced with the kind permission of CSIRO Editorial Services.)

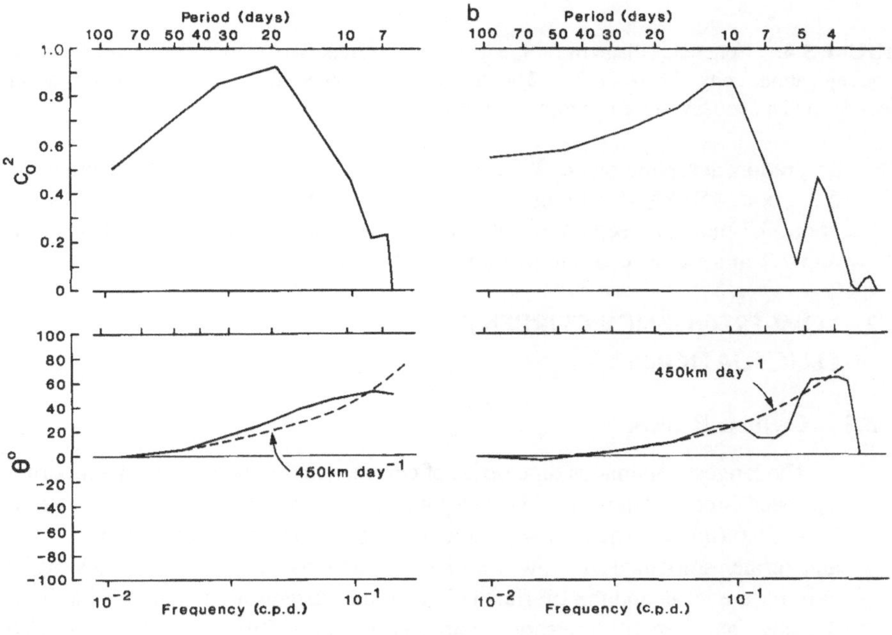

FIGURE 5.3. Coherence squared (c_o^2) and phase (Θ) relationships between (a) sea levels at Townsville and Carter Reef and (b) longshore currents near Cape Upstart and Green Island. The dashed lines indicate the theoretical phase relation for a nondispersive wave propagating longshore equatorward at 450 km d^{-1}. (From Wolanski and Pickard, 1985. Reproduced with the kind permission of Springer-Verlag.)

Barrier Reef, and 1000 km in the northern Great Barrier Reef. This implies that the central and southern regions of the Great Barrier Reef are decoupled.

The propagation speed of the LF sea-level disturbances was calculated from the difference between times of arrival for disturbances reaching pairs of locations. The propagation is northward, i.e., the disturbances propagate with the coast on their left. The speed of propagation (Figure 5.5) decreases with increasing

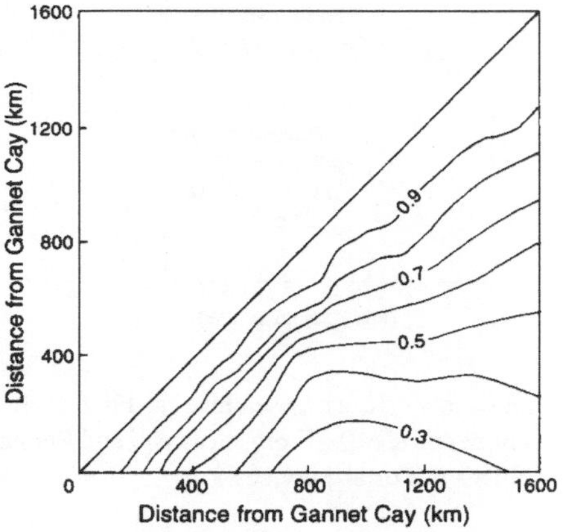

FIGURE 5.4. Coherence diagram for sea levels in the Great Barrier Reef. The diagonal is the zero separation, unity coherence line. The distance is measured along the outer Great Barrier Reef from Gannett Cay (22°S). (From Andrews, personal communication.)

separation; at separations <~200 km, it is about 4.8 m s^{-1} (the dotted line in Figure 5.5. about 450 km d^{-1}; in agreement with that inferred from Figure 5.3); it is about 3.7 m s^{-1} at a separation of 300 km, 2.8 m s^{-1} at a separation of 600 km, and 2.2 m s^{-1} at a separation of about 1200 km.

5.2 LOW-FREQUENCY CURRENT FLUCTUATIONS

5.2.1 Central Region

The longest continuous time series of currents on the shelf is that at a mooring site near Green Island (16.8°S). The longshore current component is shown in Figure 5.6 (line c). The data were smoothed to remove tidal fluctuations (line d) and further smoothed to show seasonal fluctuations (line e). In this area, tidal currents are weak and the LF fluctuations of the currents are readily visible from the raw data. The LF longshore wind component at Rib Reef (18.5°S) is also shown in Figure 5.6 (line a), together with the seasonal fluctuations (line b). Clearly, there are large temporal fluctuations of the wind and the currents. Autospectra of the LF longshore currents and wind components show, as for the sea level, that the spectra are energetic at all periods >4 d (Figure 2.4). Coherence calculations (Wolanski and Pickard, 1985) show that the LF sea-level and longshore wind components are coherent with each other for periods of about 3 to 100 d. Also, the LF longshore current at Green Island is coherent, at all periods >3 d, with the LF longshore wind component that it lags by about 1 d. The LF fluctuations of longshore currents are also coherent with each other, propagating northward at a speed of about 450 km d^{-1} (Figure 5.3b). A prevailing southward current of about 0.2 m s^{-1} was observed over 2.5 years of observations near Green Island.

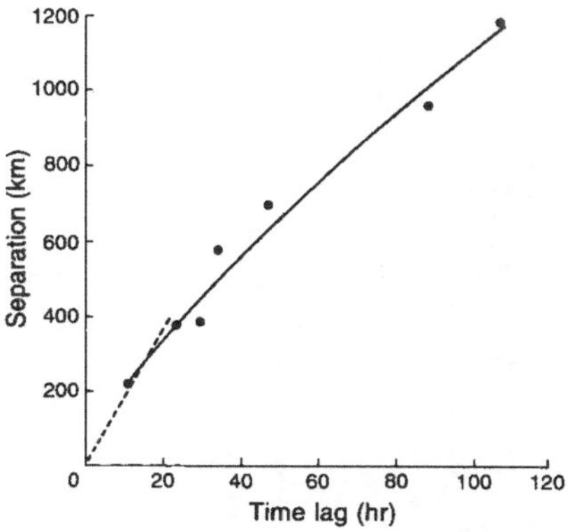

FIGURE 5.5. Time lag diagram for low-frequency sea-level fluctuations between Gannett Cay (22°S) and Raine Island (11.6°S) for all pairs of locations that showed meaningful cross-correlation and/or coherence at the 95% significance level. All propagations were northward. The dotted line is from Wolanski and Bennett (1983). (From Andrews, personal communication.)

This net southward current is due to the presence of the East Australian Current further offshore (see Section 5.2.2).

From an examination of the data, Wolanski and Bennett (1983) and Wolanski and Pickard (1985) found that the LF longshore currents v on the mid-shelf are in geostrophic balance with the cross-shelf sea level gradients, i.e.,

$$-fv = -g\frac{\partial \eta}{\partial x}$$

where f is the Coriolis parameter, η the sea level, and the x-axis is oriented across-shelf.

Near-surface currents have been measured with surface drifters by Collins and Walker (1985), who released 53,800 surface drifters (rectangular cards 11 × 31 cm, about 6% of the area emergent and exposed to the wind) at a number of sites, mainly in the central region. About 4.9% of the cards were found. The study suffers from the same questions that can be asked of all drifter studies, i.e., did those cards follow the paths typical of those of the majority (95.1%) of the cards that were not found? Collins and Walker found that in the southeast trade wind season, the cards drifted in the same direction as the wind, with a speed of 1.3 to 5.5% of the wind speed. The rest of the time the drift was more variable and in some cases was southward. The surface current of up to 5.5% of the wind speed is comparable to that measured with current metres at mid-depth, as 4% of the wind speed (Wolanski and Pickard, 1985). Wolanski and Ridd (1986) found that, in the absence of river floods, variations of the LF currents across the shelf and from the top to the bottom metre were generally <20% of the current velocity.

In the Great Barrier Reef matrix, the gathering of representative current data is complicated greatly by the presence of reefs. The LF longshore currents within

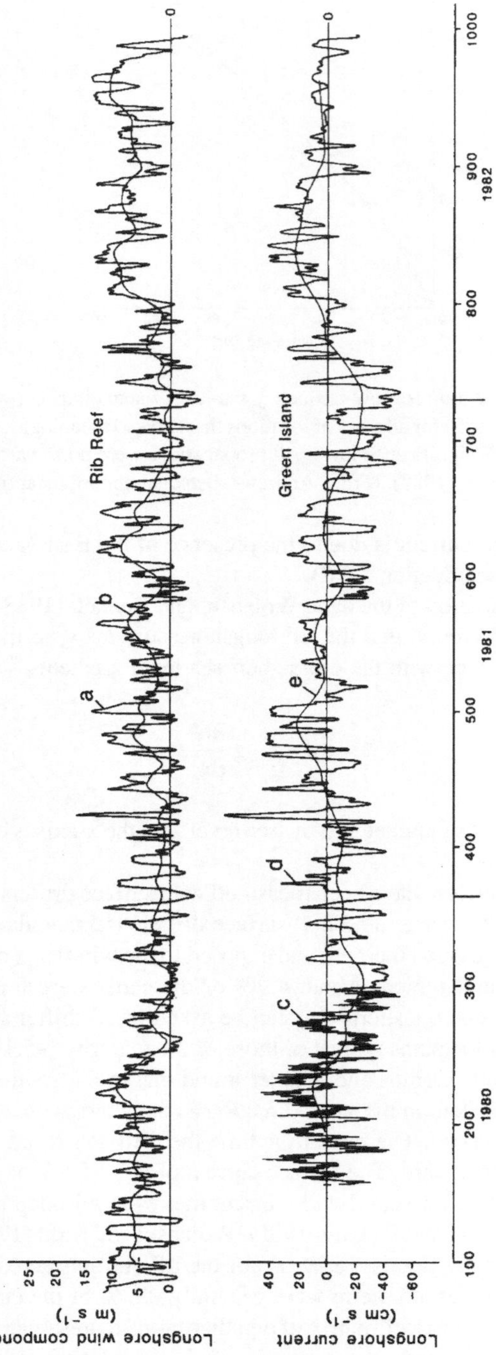

FIGURE 5.6. Time series plot of (a) the low-frequency and (b) seasonal longshore wind component at Rib Reef, and of the (c) half-hourly, (d) low-frequency, and (e) seasonal longshore current at a mooring site off Green Island. (Adapted from Wolanski and Pickard, 1985.)

Palm Passage (18.5°S), an area of least reef density, fluctuated with the wind, with a net prevailing southward current of about 0.08 to 0.2 m s^{-1} (Figure 5.7), presumably driven by the East Australian Current (Andrews, 1983b). Only 100 km to the south, in an area of higher reef density, this southward drift showed greatly spatial variability, with current-metre mooring sites a few kilometres apart experiencing very different currents, sometimes of opposite sign (Figure 4.5).

5.2.2 Coral Sea Adjoining the Central Great Barrier Reef

Just offshore from the reef matrix, at the shelf break (120-m depth) near Myrmidon Reef (18.3°S), the East Australian Current was measured as a southward current of about 0.55 m s^{-1} at a 27-m depth, decreasing to about 0.2 m s^{-1} at a 100-m depth (Figure 5.8). This current also fluctuated with the wind, the fluctuations being largest near the surface.

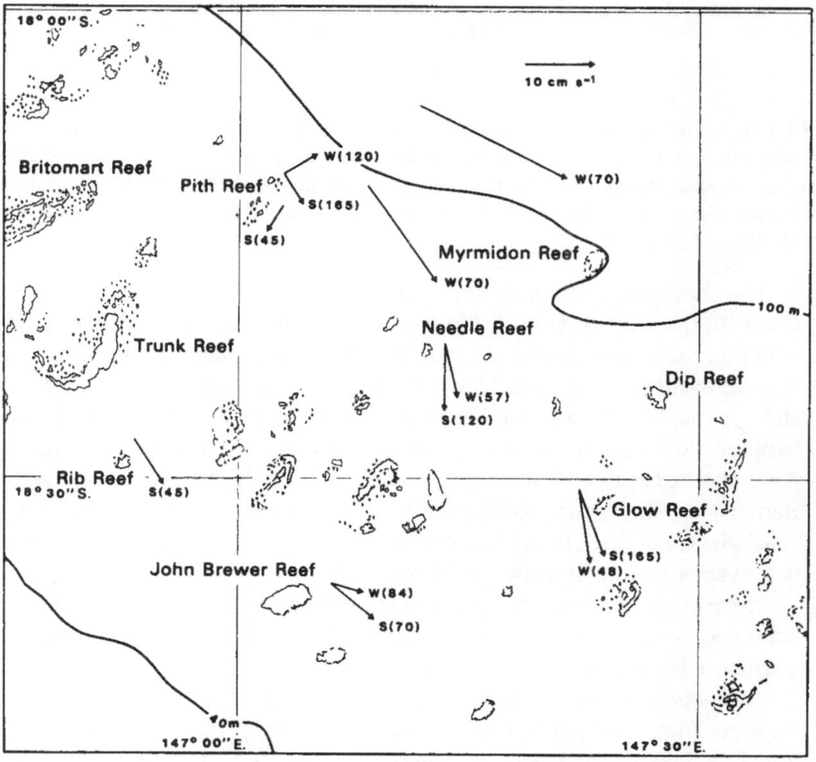

FIGURE 5.7. Seasonal mean currents in the central region in winter (w) and summer (s). The numbers in brackets refer to the record lengths in days. (From Andrews, 1983b. Reproduced with the kind permission of CSIRO Editorial Services.)

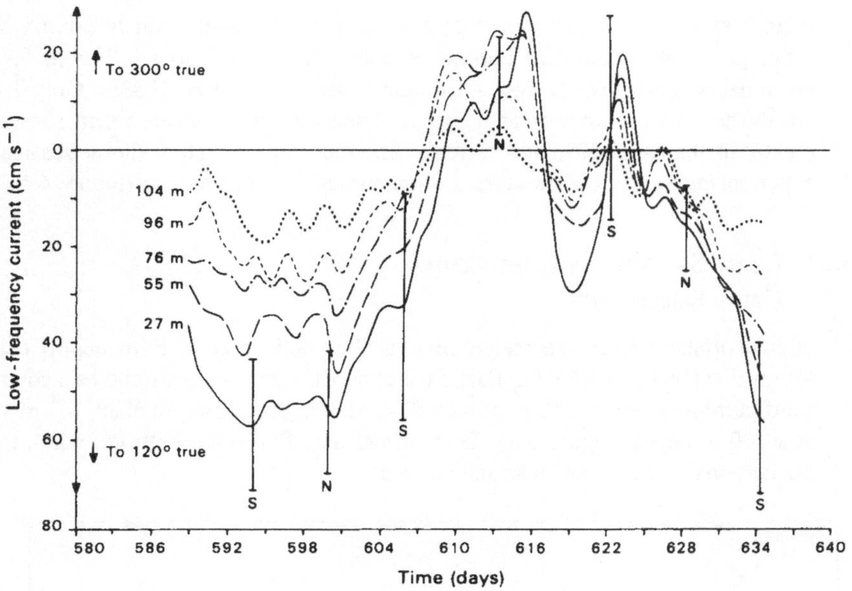

FIGURE 5.8. Time series of low-frequency longshore currents (positive if northward) at five depths at a current metre at site E near Myrmidon Reef (see location map in Figure 4.9). The vertical bars indicate the range of the tidal current fluctuations at a 27-m depth. S, spring tide; N, neap tide. (From Wolanski and Pickard, 1983. Reproduced with the kind permission of CSIRO Editorial Services.)

The geostrophic current in the western Coral Sea immediately adjacent to the Great Barrier Reef is westward in the Coral Sea toward the Great Barrier Reef, with a subsurface maximum at 150 m (Church, 1983; Andrews and Clegg, 1989). The westward-flowing current has a discharge of about 12 Sv (1 Sv = 10^6 m^3 s^{-1}) and is created by the westward baroclinic influx of the South Equatorial Current between the Solomon Islands and Vanuatu. This current splits up at a bifurcation point and exits as two longshore currents, each of about 6 Sv, along the Great Barrier Reef slope, one southward (the East Australian Current) and one northward (the Coral Sea Coastal Current; Figure 5.9). The location of the bifurcation point varies seasonally between 18 and 14°S.

There is also evidence of a permanent undercurrent flowing northward on the outer continental shelf of the Great Barrier Reef and under the East Australian Current (Church and Boland, 1984).

Fluctuations in the speed of the East Australian Current on the continental slope generate oscillations of temperature and salinity (Figure 5.10) that were first found by Andrews and Gentien (1982). These oscillations appear to be internal waves, sloshing up and down the continental slope, and trapped along the continental slope within a Rossby radius of deformation. They occasionally spill subthermocline nutrient-rich water onto the shelf as near-bottom, cold water moving cross-shelf across the reef matrix. The typical intrusion layer thickness was about 5 m initially before extending by mixing to half depth, and typical

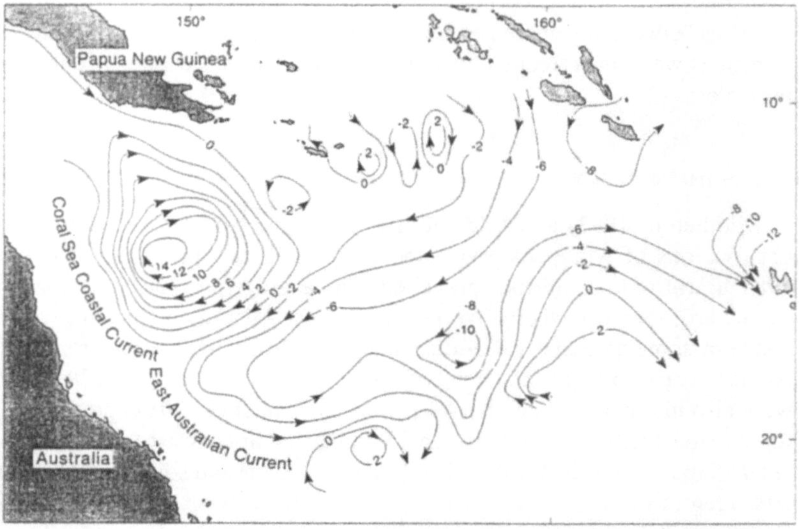

FIGURE 5.9. Contours of volume transport for the top 1000 m at 2.5-Sv intervals, for the Western Coral Sea. (Adapted from Andrews and Clegg, 1989.)

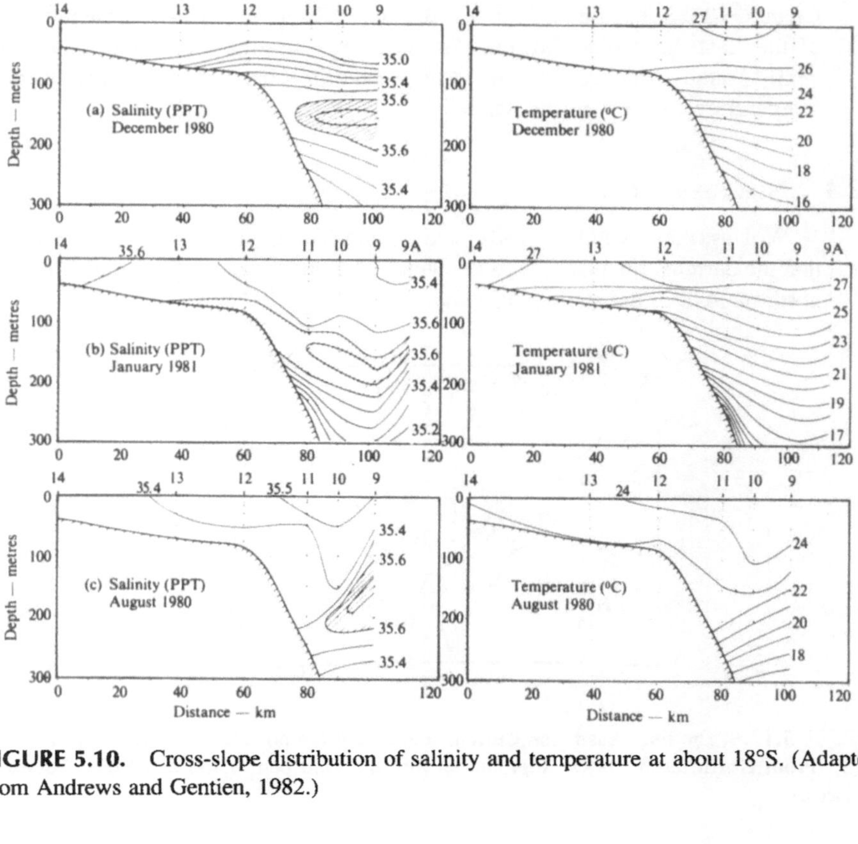

FIGURE 5.10. Cross-slope distribution of salinity and temperature at about 18°S. (Adapted from Andrews and Gentien, 1982.)

contrasts between the upper and lower layers were 3 to 4°C and 0.25 psu. These intrusions were most frequent in summer, when the mixed layer is thinner than in winter.

5.2.3 Southern Region

Middleton (1983) and Middleton and Cunningham (1984) reported field observations of wind, sea level, and currents from April 1980 to February 1981. Barotropic conditions prevailed. The average currents were longshore southward, forced further offshore by the East Australian Current, with evidence of some seasonal variations, as the southward drift was stronger after August. The currents fluctuated with periods of 7 to 20 d. These fluctuations were also highly correlated between sites. Griffin et al. (1987) presented data for six sites further south, between 23 and 24°S. In contrast to the southward net drift in the central (Wolanski and Pickard, 1985) and southern (Middleton, 1983) regions of the Great Barrier Reef, a northwestward drift was observed, peaking at 0.3 m s^{-1}. This flow was in bursts of 6- to 10-d periods. The sea-level fluctuations were in geostrophic balance with the longshore currents. They postulated the existence of a clockwise rotating eddy in the Capricorn Channel (23°S; Figure 5.11) that appears and disappears with the meandering of the East Australian Current. W. Skirving (personal communication) recently found evidence for this eddy in satellite observations of sea surface temperature in Capricorn Channel.

5.2.4 Northern Region

Wolanski and Ruddick (1981) and Wolanski and Thomson (1984) reported that the currents fluctuated with the wind, with no evidence of a southward drift north of 14°S in the absence of wind.

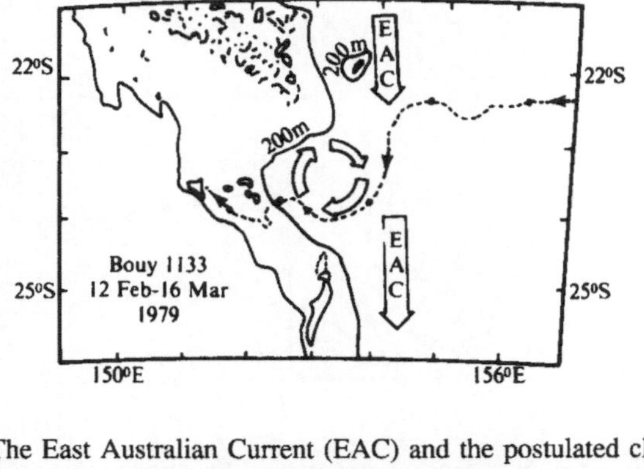

FIGURE 5.11. The East Australian Current (EAC) and the postulated clockwise eddy near 28°S. (From Griffin et al., 1987. Reproduced with the kind permission of CSIRO Editorial Services.)

The LF currents were strongly dependent on, and could not be analyzed independently from the tidal currents, being strongest in regions where the tidal currents were weakest (Figure 5.12).

Current-metre data near Pearce Cay (9.5°S) in the Great North East Channel for 5 months in 1981 to 1982 (Wolanski and Thomson, 1984) and for 9 months in 1991 to 1992 (Wolanski, unpublished data) show the presence of alternating northward and southward longshore LF currents (Figure 5.13). the flow is topographically steered by the Warrior Reefs to the west and by a dense matrix of reefs to the east (Figure 3.18).

Current-metre data from sites in Torres Strait show periods of alternating eastward (from the Gulf of Carpentaria to the Great Barrier Reef) and westward currents (Figure

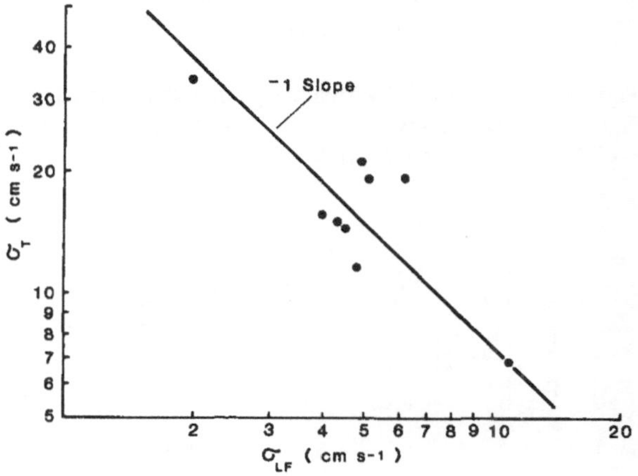

FIGURE 5.12. Plot of the standard deviation of low-frequency currents (σ_{LF}) vs. that of the tidal currents (σ_T) at mooring sites in the northern region. (From Wolanski and Thomson, 1984. Reproduced with the kind permission of Academic Press.)

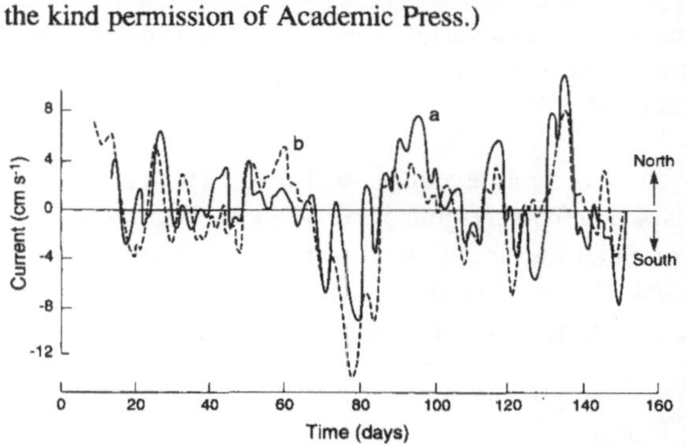

FIGURE 5.13. (a) Observed and (b) computed low-frequency currents near Pearce Cay (9.55°) in the Great North East Channel. Location map in Figure 1.2d. Time is expressed in day no. from November 10, 1981.

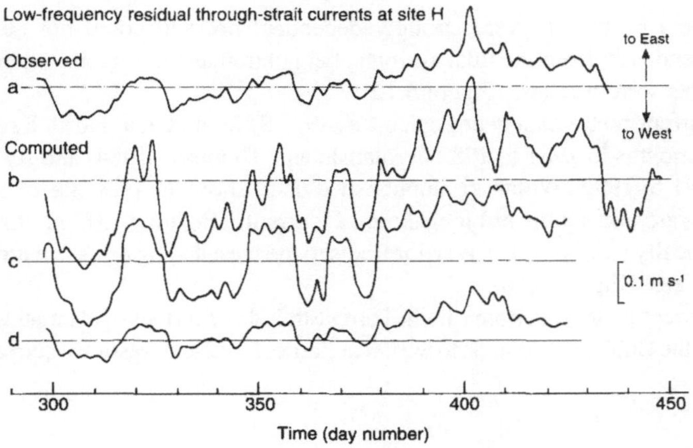

FIGURE 5.14. Time series plot of the low-frequency currents through Torres Strait at site H (location map in Figure 1.2d), (a) observed and (b to d) computed under various assumptions described in the text. The time is expressed in day no. in 1986/1987.

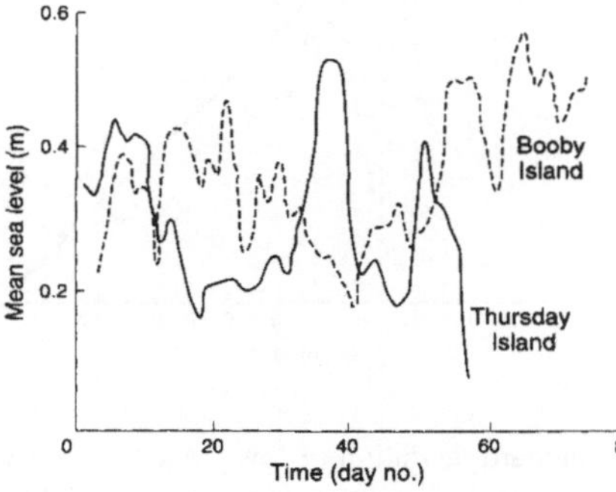

FIGURE 5.15. Time series plot of the low-frequency sea level on either side of Torres Strait at Thursday Island and at Booby Island (see location map in Figure 1.2a). Time is expressed in day no. from September 30, 1979.

5.14, line a). These currents are small, <0.1 m s^{-1}. There are also large LF sea-level oscillations, typically 0.3 m (Figure 5.15). These oscillations are incoherent on either side of Torres Strait (Amin, 1978; Wolanski and Ruddick, 1981; Wolanski and Thomson, 1984; Wolanski et al., 1988b). These resulted in fluctuations of up to 0.6 m of the LF sea-level difference on either side of Torres Strait.

5.2.5 Gulf of Papua

Data from a cross-shelf transect of current metres revealed that the circulation is weak and variable (Wolanski et al., 1992c). The circulation is controlled by several factors operating simultaneously: (1) the eastward Coral Sea Coastal Current in the northwest Coral Sea (Figure 5.9), (2) the wind, (3) the river runoff which results in a strong salinity stratification of the Gulf, and (4) the current through the Great North East Channel (Figure 5.13). The residence time of the brackish water is 1 to 2 months. Little of the brackish water leaves the Gulf through Torres Strait.

6 Models of the Low-Frequency Circulation

As described in Chapter 5, the low-frequency (LF) fluctuations of sea level, currents, and wind in the southern and central regions of the Great Barrier Reef are coherent with each other. The fluctuations travel northward. They appear to be trapped over the shelf. These observations suggest that the fluctuations are due to continental shelf waves.

Continental shelf waves are long waves controlled by the earth's rotation, and the topography, and in their most common mode are generated by the wind. In the Great Barrier Reef they are trapped on the continental shelf and move northward with the coast on their left. As summarized by Griffin and Middleton (1986), these waves have several modes. In this first mode (barotropic), they take the form of long waves with maximum amplitude at the coast and zero amplitude in the deep ocean. The circulation is longshore, alternating up and down the coast and accompanied by cross-shelf motions with smaller velocities, resulting in a series of cells (Figure 6.1a). In higher modes, baroclinic motions result in wave-like vertical displacements of the thermocline intersecting the continental shelf slope (Figure 6.1b). The vertical oscillations of the thermocline may be tens of metres, while the vertical displacement of the free surface may be only 0.3 m.

In the central region of the Great Barrier Reef, the shelf width L is about 100 km and the mean depth H is 40 m. For wind fluctuations with a period of 5 to 10 d, first-mode waves have a wave length of about 4000 km, which is much longer than the length of the Great Barrier Reef.

6.1 BAROTROPIC CURRENTS IN THE CENTRAL AND SOUTHERN REGIONS

The linearized equations of motion for barotropic flow over a shallow shelf bounded by a deep ocean are (Middleton, 1983; Middleton and Cunningham, 1984)

$$\frac{\partial u}{\partial t} - fv = -g\frac{\partial \eta}{\partial x} + (\tau_w^x - \tau_b^x)/\rho h_s \tag{6.1}$$

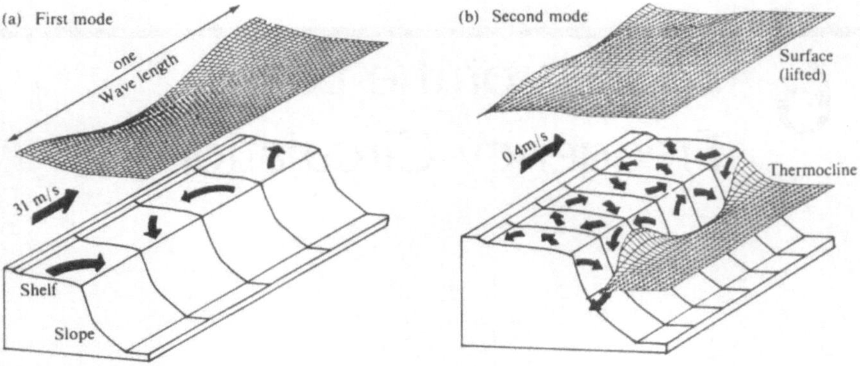

FIGURE 6.1. Sketch of the circulation generated by a trapped continental shelf wave. (a) Mode 1. (b) mode 2. (From Griffin and Middleton, 1986. Reproduced with the kind permission of the American Meteorological Society.)

$$\frac{\partial v}{\partial t} + fu = -g\frac{\partial \eta}{\partial y} + (\tau_w^y - \tau_b^y)/\rho h_s \tag{6.2}$$

$$\frac{\partial \eta}{\partial t} + \partial(hu)/\partial x + \partial(hv)/\partial y = 0 \tag{6.3}$$

where (u, v) are the depth-averaged velocities along the (x, y) axes oriented respectively to the east cross-shelf and the north longshore, η is the free-surface elevation, h is the depth, f is the Coriolis parameter, g is the acceleration due to gravity, τ_w is the wind stress, τ_b is the bottom friction stress, and ρ is the density.

Analytical solutions of these equations are available, provided a number of simplifying assumptions are made (Middleton, 1983). First (Figure 6.2a), the

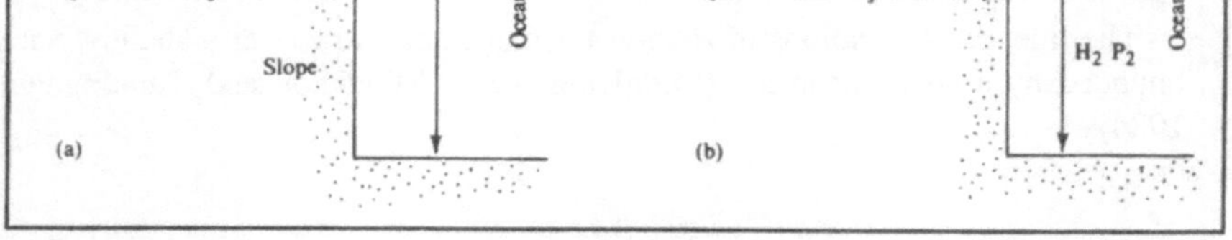

FIGURE 6.2. Shelf wave model geometry following Middleton (1983).

shelf is assumed to be flat (h_s = constant) and the continental slope vertical, the ocean having a depth H. Second, only the longshore wind component is considered. Third, the wind stress is uniform on the shelf. Fourth, the longshore current is assumed to be in geostrophic balance with the cross-shelf sea-level slope, so that Equation 6.1 simplifies to

$$fv = g\,\partial\eta/\partial x \tag{6.4}$$

Fifth, Equation 6.3 can be simplified by making the rigid lid assumption (Gill and Schumann, 1974), i.e., assuming that $\partial\eta/\partial t$ is negligible, and becomes

$$\partial(h_s u)/\partial x + \partial(h_s v)/\partial y = 0 \tag{6.5}$$

Wolanski and Bennett (1983) verified experimentally the validity of Equations 6.4 and 6.5. Equation 6.5 implies nondivergence, so, for a constant h_s, u and v can be expressed in terms of a stream function ψ,

$$u = \frac{1}{h_s}\frac{\partial\psi}{\partial y} \text{ and } v = -\frac{1}{h_s}\frac{\partial\psi}{\partial y} \tag{6.6}$$

Sixth, the longshore bottom stress term is linearized,

$$\tau_b = ru \tag{6.7}$$

where r is a friction coefficient.

This leads to the forced-wave equation (Middleton, 1983; Middleton and Cunningham, 1984)

$$\frac{1}{c}\frac{\partial\psi}{\partial t} + \frac{\partial\psi}{\partial x} + \frac{r\psi}{ch_s} = \tau_w^y \ \rho c \tag{6.8}$$

where c is the free-wave speed,

$$c = -f L_s(1 - h_s/H_o) \tag{6.9}$$

where L_s is the width of the shelf. Estimates of c vary between 8.7 m s^{-1} (750 km d^{-1}) in the southern zone (Middleton and Cunningham, 1984) and 2.5 m s^{-1} (217 km d^{-1}) in the central region of the Great Barrier Reef (Wolanski and Bennett, 1983).

To obtain simple analytical solutions, further assumptions are needed. A periodic wind stress is assumed,

$$\tau_w^y = \tau_o \cos(ky - \omega t) \tag{6.10}$$

where $\omega = 2\pi/T$ is the frequency, T is the period, τ_o is the intensity, and $k = 2\pi/L_o$, where L_o is the dimension of the wind field. The wind stress field in this representation moves northward at a speed ω/k. Typically in the southeast trade wind season, T = 8 to 20 d and L_o = 9200 km. This representation of the wind stress neglects the fact that the wind stress does not reverse sign, but is always positive longshore northward in the southeast trade wind season.

A closed-form solution can be found, consisting of free and forced waves. Forced waves are uniform in amplitude over the shelf because τ_o is assumed constant, and move at the speed ω/k_o of the wind stress. The free waves move northward at a speed c. However, the free waves are damped by bottom friction and decrease with increasing distance from the origin at y = 0 following an exponential damping term $\exp(-ry/cH_s)$. The longshore current is found to be (Middleton and Cunningham, 1984)

$$v = G\,\tau_o / \rho c\, h_s \left[\gamma^2 + \left(ko - \frac{\omega}{c}\right)\right]^{1/2} \tag{6.11}$$

where $\gamma = \rho/c\,h_s$, G is a function of ω, ko, and γ, and θ and δ are trigonometric functions of k, ω, y, and c. The solution is simplest near the origin, i.e., for $ry/ch_s \ll 1$, where v is a linear function of y.

A particularly simple solution results when the wind stress is spatially uniform (k = 0)

$$\tau = \tau_o\, e^{i\omega\tau} \tag{6.12}$$

leading to the simple closed-form solution for the longshore current (Wolanski and Bennett, 1983)

$$v = \frac{\tau_o}{h_s}\, e^{i\omega t}\left[1 - e^{-i\omega y/c} e^{-ry/h_s}\right] / i(\omega - ir/h_s) \tag{6.13}$$

This implies that sea-level and current disturbances propagate longshore with a phase speed of 2c independent of the friction. However, the amplitude of v and η depends upon the distance y.

This simple analytical model captures some key features of the dynamics of long waves in the central region of the Great Barrier Reef. In the central region, the predictions of the speed of northward propagation of the LF sea-level and current fluctuations agree with the observations (450 km d^{-1}, Figure 5.3); η is largest at the coast, the long waves are trapped on the shelf (Figure 5.2), v and η are in phase, the cross-shelf velocities are much smaller than the longshore velocities, and v varies little across the inner and mid-shelf (Wolanski and Bennett, 1983). Assuming that the ill-defined southern boundary (y = 0) is at Mackay (21.2°S), from the northward decay of the long waves, Wolanski and Bennett (1983) found that $r = 1$ to 2×10^{-4}m s^{-1}. The field data showed a small lag between oceanic disturbances at the coast and at the shelf break, which the model does not reproduce and which presumably is due to the slope of the sea floor (Brink and Allen, 1978).

In the southern region, Middleton and Cunningham (1984) assumed that y = 0 at 23°S, at the mouth of the Capricorn Channel where the shelf width increases abruptly with decreasing latitude. They obtained a best fit for $L_s = 180$ km, $h_s/H = 35/500$, and $r = 2.5 \times 10^{-4}$m s^{-1}. A further test of the model was undertaken by Griffin and Middleton (1986) in the region between 23 and 25°S, a region bounded to the south by Fraser Island, which extends practically to the shelf break. This region was chosen to test Middleton and Cunningham's (1984) model of wind-driven shelf waves with a known geographic origin, as it was believed that Fraser Island (25°S) would effectively block shelf waves originating further south. While the model could reproduce some of the observations, there were also events which were not generated by the local wind, a finding suggesting that free waves from further south and of higher mode could bypass Fraser Island and enter this area. Griffin and Middleton (1986) also suggested that the widening of the shelf at the Capricorn Channel reduces the speed of free waves and results in deposition of sand, thereby helping to maintain the supply of sand to Fraser Island.

6.2 INFLUENCE OF THE REEF ON CROSS-SHELF MOVEMENTS

In the central region, the Great Barrier Reef appears transparent to shelf waves. Its only effect, then, is to reduce by a distance d the effective shelf width exposed to longshore currents (Figure 6.2), so that Equation 6.9 becomes

$$c = -f(L_s - d)(1 - h_s/H) \tag{6.14}$$

To calculate the degree of transparency of the Great Barrier Reef to shelf waves, Middleton (1983) presented an analytical model. He assumed that only free waves are present and that there is no friction over the shelf (r = 0). A two-layer ocean was assumed (Figure 6.2). The layers have a thickness and density of H_1, ρ_1 and H_2, ρ_2, respectively. The Great Barrier Reef was modeled as a submerged barrier of depth h_r ($<h_s$), and of width d. The flow over the model reef was controlled by a linear friction. Over the model reef, Equation 6.3 remains unchanged, but Equations 6.1 and 6.2 simplify to

$$fv = g\,\partial\eta/\partial x + ru/h_r \tag{6.15}$$

$$fu = -rv/h_r \tag{6.16}$$

The sea-level difference $\Delta\eta$ across the reef was calculated by Middleton (1983) to be

$$\Delta\eta = h_s u(r\,d/gh_r^{\,2})(1 + f\,h_r^{\,2}/r^2) = h_s u\,F \tag{6.17}$$

where F is the "reef opacity factor". This model suggests that the model reef has little effect on Kelvin waves and northward traveling baroclinic shelf waves, but significantly increases the speed c of barotropic shelf waves, the speed increasing with decreasing period. For instance, in the presence of the model reef, the wave speed is 8.75 m s^{-1} for waves of period >20 d, but is 14.6 m s^{-1} for waves of a 4-d period. There is some support for these predictions in Middleton's data set for the Mackay region after removing the directly wind-forced components from the data.

Wolanski and Bennett (1983) modeled the Great Barrier Reef as a barrier along the shelf break completely blocking the exchange of water between the shelf and the ocean except through a number of gaps, of depth equal to that of the shelf. These gaps occupy a fraction of the length of the shelf. The authors assumed a linear friction law both on the shelf and in the passages. They found that in the central region where reef density is least, friction over the shelf and in the reef passages leads to an e-folding decay time of about 2 and 10 d, respectively. This implies that in the central region of the Great Barrier Reef, the effect of friction in reef passages can be discarded and that the reef is transparent to shelf waves. This finding runs somewhat opposite to, but does not contradict, that of Middleton (1983). The reason for the difference appears to be the different geometry. In the southern region, shelf water is deeper so that friction on the shelf is less important, and the reef density and reef width are much higher, so that blockage of the cross-shelf flow is more pronounced than in the central region (Huthnance, 1985).

6.3 BAROCLINIC EFFECTS IN THE CENTRAL REGION

The LF oscillations of the thermocline along the continental slope of the central region appear to be confined to the proximity of the shelf break (Figure 5.10). This finding suggests that these oscillations may be low-frequency baroclinic waves trapped along the continental slope, possible mode 2 shelf waves (Figure 6.2). This was modeled analytically by Wolanski (1986e). He assumed a vertically homogeneous, flat continental shelf, of depth h_s and width L (Figure 6.3), a vertical continental slope, and a two-layer ocean, with upper and lower layers of respective depth and density h and h′ and ρ and ρ'. The depth of the mixed layer

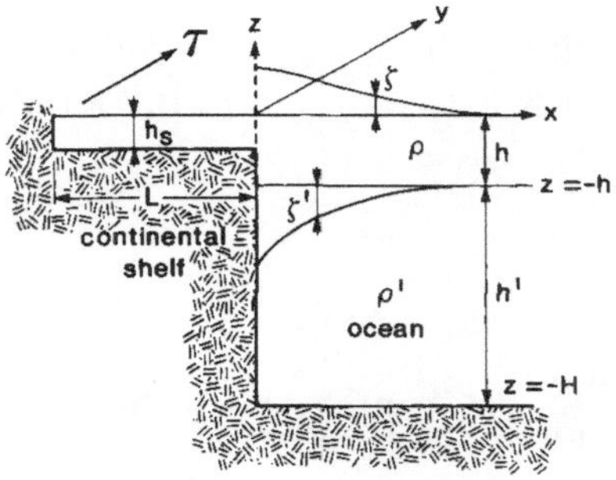

FIGURE 6.3. Baroclinic slope wave model geometry.

is h. At rest, $h > h_s$, the thermocline is located offshore from the shelf break. These assumptions are more or less representative of the geography and thermal structures in 1983 to 1984, when a mooring was maintained for a year on the shelf slope with current metres both above and below the thermocline. Longshore currents, but no longshore gradients, were allowed. A periodic, spatially uniform, wind stress was applied,

$$\tau = \rho\, u_*^2\, e^{i\omega\tau} \tag{6.18}$$

where u_* is the shear velocity. A linear friction law was assumed over the shelf. Bottom friction was neglected in the ocean because of its great depths. Closed-form, trapped-wave solutions (decaying exponentially to zero at $x = \infty$ in the ocean) apply for the shelf (Equation 6.13) and in the coastal ocean, for the upper layer

$$u = i\,\alpha\, e^{i\omega t}\, e^{l_2 x}/h \tag{6.19}$$

$$v = f\,\frac{\alpha\, e^{i\omega t}}{\omega}\left[e^{lx} + e^{l_2 x}\right]/h \tag{6.20}$$

and for the lower layer

$$u' = -\frac{i\,\alpha\, e^{i\omega t}}{\omega}\left[e^{lx} - e^{l_2 x}\right]/h' \tag{6.21}$$

$$v' = -\frac{f\,\alpha\, e^{i\omega t}}{\omega}\left[e^{lx} - e^{-l_2 x}\right]/h' \tag{6.22}$$

and the thermocline displacement is given by

$$\zeta' = \frac{\beta(h+h')}{\varepsilon\, h'\, c_o\, l_2} e^{l_2 x} e^{i\omega t} + \frac{\beta\, e^{lx}\, e^{i\omega t}}{\left\{1+\left[h'/(h+h')\right]c^2\right\}} \tag{6.23}$$

where

$$\alpha = \frac{L\, u_*^2\, \omega}{c_s(r+i\omega)} e^{-i\omega y/c_s} e^{-ry/c_s}, \quad l = -\left[\left(f^2-\omega^2\right)/c^2\right]^{1/2} < 0$$

$$l_2 = -\left(f^2-\omega^2\right)^{1/2} / \left[g\varepsilon h\left(h+h'\right)/h\right] \tag{6.24}$$

where

$$c_2^2 = g\varepsilon h'h/h+h', \quad \varepsilon = \frac{\rho'-\rho}{\rho} << 1, \quad c_o^2 = gH, \text{ and } c_s^2 = g\, h_s \tag{6.25}$$

The model qualitatively reproduces a number of observations of currents on the continental slope. It implies that under the action of a reversing longshore wind, reversing longshore currents result in the upper ocean with little lag in the top layer. The thermocline is also alternatively raised and lowered as an internal wave sloshing up and down the continental slope and trapped along the slope (i.e., decaying to zero offshore from the shelf break). The field observations suggest maximum vertical amplitudes of the displacement of the thermocline of 50 m (Wolanski, 1986e), implying that the internal waves can readily spill into the shelf. The model suggests that such intrusions would be extensive in the longshore direction, a prediction that is consistent with the observations of Andrews and Furnas (1986).

6.4 NORTHERN REGION

The analytical models of shelf waves are successful in describing some of the physics of the LF water circulation in the southern and central regions of the Great Barrier Reef. These models require a number of simplifying assumptions, including a greatly simplified geometry and neglect of the interaction of LF and tidal currents in shallow waters. These assumptions seem quite unrealistic in the northern region. This region is shallower, the flow is encumbered by a large number of coral reefs and submerged sand banks are scattered densely on the shelf, and reef-free waters are found only in two narrow (a few kilometres wide) channels, one near the coast and one near the outer reef. Bottom frictional effects thus are dominant. The time-derivative terms in Equations 6.1 to 6.3 become negligible and the equations of motion on the shelf are (Wolanski and Thomson, 1984)

$$-f\,v = -g\frac{\partial \eta}{\partial x} + \frac{\tau_w^x}{h} \tag{6.26}$$

$$fu = -g\frac{\partial \eta}{\partial y} + \frac{\tau_w^y}{h} - \frac{\tau_b^y}{h} \tag{6.27}$$

$$\frac{\partial hu}{\partial x} + \frac{\partial hv}{\partial y} = 0 \tag{6.28}$$

These equations reflect a simple balance between the wind stress, the bottom friction, and the sea surface slope. Free waves do not exist. Any northward advection of sea levels and current fluctuations can only be caused by a northward advection of the wind stress field. Closed-form analytical solutions exist in the form of "arrested topographic waves" (e.g., Csanady, 1978, 1982). These solutions suggest that for a long, flat continental shelf, the longshore currents are coherent with the longshore wind component but independent of both the distance from the shore and the cross-shelf wind, and that the zone of return flow is located past the shelf break. Field data (Wolanski and Thomson, 1984) generally confirm these predictions but show also that the amplitude of the LF current fluctuations varies spatially and is inversely correlated to the amplitude of the tidal currents (Figure 5.12). This finding suggests a nonlinear interaction, through the bottom friction effect, between strong tidal currents and small LF currents (Csanady, 1976; Winant and Beardsley, 1978; Provis and Lennon, 1983). Thus,

$$\tau_b = r\,v \tag{6.29}$$

where r, the bottom friction coefficient for LF motions, is not constant but varies spatially following

$$r = b\,\sigma_T \tag{6.30}$$

where b is a constant and σ_T is the standard deviation of the tidal current (Wolanski and Thomson, 1984).

Thus, LF and tidal currents cannot be modeled independently. The analytical model still yields useful results in limited areas where the spatial gradients of tidal currents are small, such as in the Great North East Channel (Figure 3.18). Simplifying assumptions are necessary. The geometry is simplified (Figure 6.4) so that the LF currents are confined to a flat narrow channel between Torres Strait (the Warrior Reef) to the west and the dense matrix of reefs to the east, with width B and length L_1. The shelf width is L, where $B/L \ll 1$. Torres Strait is assumed to be impermeable to long waves. The sea level in the Gulf of Papua

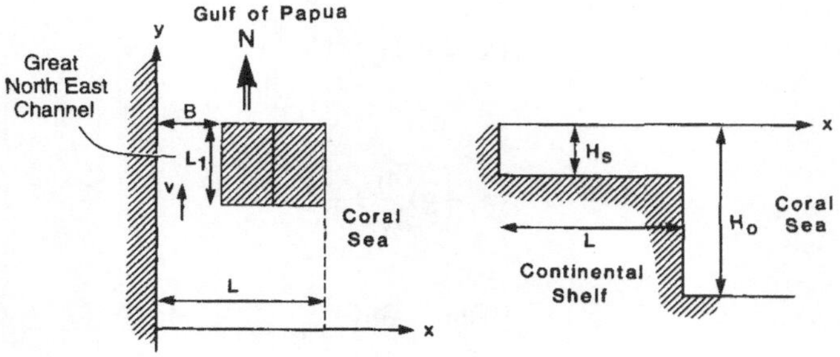

FIGURE 6.4. Model geometry for the Great North East Channel.

is assumed to be equal to that of the Coral Sea. Finally, a nonlinear bottom friction law is used

$$\tau_b = C_d |v| v \tag{6.31}$$

where C_d is a drag coefficient. The wind drives a longshore current v through the Great North East Channel two ways: (1) by the along-Channel wind stress along the axis of the Channel and (2) by the cross-shelf wind component piling up water on the coast. This piling up of water generates a longshore sea-level gradient along the Channel. Equations 6.26 to 6.28 yield

$$v = \pm \sqrt{\left[\frac{\tau_w^y - a\,\tau_w^x}{C_d} + |v_o| v_o \right]} \tag{6.32}$$

where v has the sign of the quantity in square brackets, $a = L/L_1 = 5$, and $v_o = 0.05$ m s^{-1} is a mean current driven by the Coral Sea Coastal Current. The 4-month-long time series of observed and predicted (using $C_d = 5 \times 10^{-2}$) longshore currents in the Great North East Channel are shown in Figure 5.13. The comparison is encouraging, suggesting that the arrested topographic wave model is successful in capturing the essential dynamics of the LF circulation.

6.5 TORRES STRAIT

The nonlinear relationship between LF and tidal currents is very strong in Torres Strait, as the tidal currents are swift and the waters are shallow (average depth $\simeq$ 10 m). The apparent friction coefficient for LF currents can be accurately estimated because field data are available for sea levels and currents, at both tidal and low frequencies, on both sides of and within Torres Strait. Field data (see Chapter 3) suggest that the dominant tidal and LF currents are oriented through Torres Strait, which can be modeled by the one-dimensional channel model shown in Figure 4.8. The equations of motion are the open-channel flow equations,

$$\partial A/\partial t + \partial Q/\partial x = 0 \tag{6.33}$$

$$\partial Q/\partial t + \partial(Q^2/A)/\partial x + gA\,\partial\eta/\partial x + g\,A\,S_f - \tau_w\,W = 0 \tag{6.34}$$

where Q is the discharge in a cross-section of area A (of width W and depth H) of the channel and S_f is the frictional slope computed from open-channel flow theory,

$$S_f = n^2 Q|Q|\,W^{4/3}/A^{10/13}$$

where n is the Manning roughness coefficient. The advantage of this formulation is that the coefficient n depends on the bottom roughness only, and is independent of the strength of the currents, whether tidal or low frequency. For a sandy bottom, $n \approx 0.025$.

The values of W and H in the respective regions were taken to be the same as those used in the tidal model (see Section 4.2.4). The wind stress was calculated from the local wind data at Thursday Island. The model is forced by the local sea levels on either side of Torres Strait. If only the LF components are used in the open-boundary conditions, the momentum balance reduces to that of an arrested topographic wave. The predicted LF currents are shown in Figure 5.14 using as open-boundary conditions on either side of Torres Strait either (line b) a constant sea level or (line c) the observed LF sea-level fluctuations. These predictions do not reproduce the observations. This suggests that the bulk friction used in these simulations is too small. The friction parameter n in this representation is uniquely determined by bottom roughness and cannot be modified to fit the data. The apparent friction for LF currents can only be increased by including the tidal stress. This was done simply by forcing the model on either side of Torres Strait by observed raw sea-level data, which thus includes both the tidal and LF currents. The resulting dominant currents are tidal. The predicted LF through-strait currents are obtained by low-pass filtering the predicted current time series and are shown in Figure 5.14 (line d). The comparison between observed (line a) and predicted (line d) through-Strait currents is encouraging. Hence, in Torres Strait, the tidal stress increases by nearly an order of magnitude the value of the bulk friction coefficient for LF currents. Clearly, therefore, tidal and LF currents are not independent in the presence of strong tidal currents in shallow water.

6.6 INFLUENCE OF THE EAST AUSTRALIAN CURRENT

King and Wolanski (1990) used a finite-difference, nonlinear, numerical model to study the depth-averaged, two-dimensional circulation in the central region of the Great Barrier Reef under the influence of the East Australian Current. The model domain, shown in Figure 6.5a, extended from the shore to the

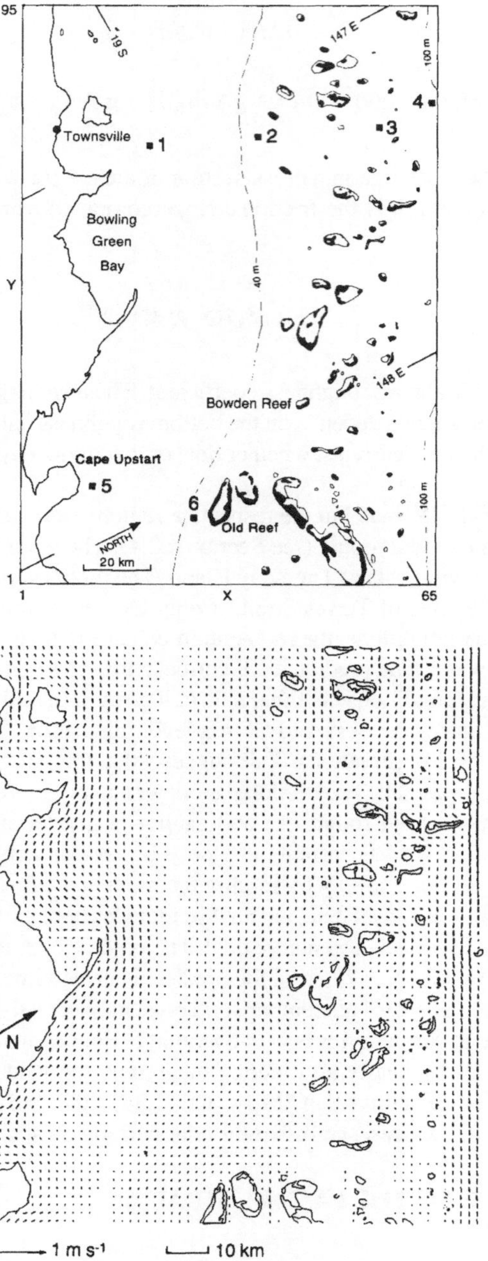

FIGURE 6.5. (a) Model domain and (b) predicted depth-averaged steady circulation under the influence of the East Australian Current and a 10 m s^{-1} southeasterly trade wind. (Adapted from Wolanski, 1992a.)

shelf slope, and the mesh size was 2031 m. The influence of the East Australian Current, and its modification by the longshore wind stress, was incorporated by including a longshore pressure gradient on the seaward open boundary. The model was calibrated against field data from six current metres (sites 1 to 6), two metres (sites 5 and 6 from Wolanski and Bennett, 1983) located on the inner shelf between Cape Upstart and Old Reef and the other four metres (sites 1 to 4) from Burrage et al. (1991) located along a cross-shelf transect from Townsville to the shelf break. The model suggests that, in moderate 10- to 15-knot southeast trade winds, the East Australian Current generates a southward drift on the mid- to outer shelf while the wind generates a northward drift in shallow coastal waters (Figure 6.5b). The decoupling of the currents across the shelf is highlighted by a zone with zero currents.

6.6.1 "Sticky" Waters

The reef density at mooring sites 1 to 4 (Figure 6.5a) is the smallest for the whole Great Barrier Reef. It is thus not surprising that in this area a southward current is present, more or less insensitive to the Great Barrier Reef, as a result of the longshore sea-level gradient generated by the East Australian Current (Middleton, 1987). Around Bowden Reef (19°S), in the southern region of the model domain shown in Figure 6.5a, current-metre data suggest that there is a nonlinear interaction between tidal currents and LF currents, as in Torres Strait. This interaction leads to the phenomenon of "sticky waters", resulting in the blocking of the East Australian Current at spring tides in an area of high reef density.

As exemplified by the case of Torres Strait described in Section 6.5, the apparent friction coefficient for LF currents is high in areas of strong tidal currents in shallow waters. In areas of high reef density where tidal currents are strong, the LF currents will be smaller than in areas of low reef density, for the same forcing by the wind and the East Australian Current. As a result, in the Great Barrier Reef, the net currents driven by a longshore pressure gradient tend to bifurcate around and avoid the areas of high reef density.

There is indeed evidence that in areas of high reef density, the intrusion of the East Australian Current can be significantly lessened. First, current-metre data at Glow and Needle Reefs (Figure 5.7) show a veering of the mean current toward the coast, suggesting that the area of high, reef density further to the southeast blocks the southward drift and deflects the mean current shoreward. Second, current-metre data near Bowden Reef (19°S) show that the LF currents fluctuate in both direction and intensity (Figure 4.5). This fluctuation appears to be modulated by the spring-neap tidal cycle, in addition to being wind dependent. Indeed, at spring tides, the mean current is smallest and oriented toward the coast instead of longshore. This blocking effect is also apparent from the trajectory of 2×2 m radar-tracked drogues tagging the water near the surface and at a 10-m depth. At spring tides on November 14 and 15, 1986, the drogues (Figure 6.6a) circulated

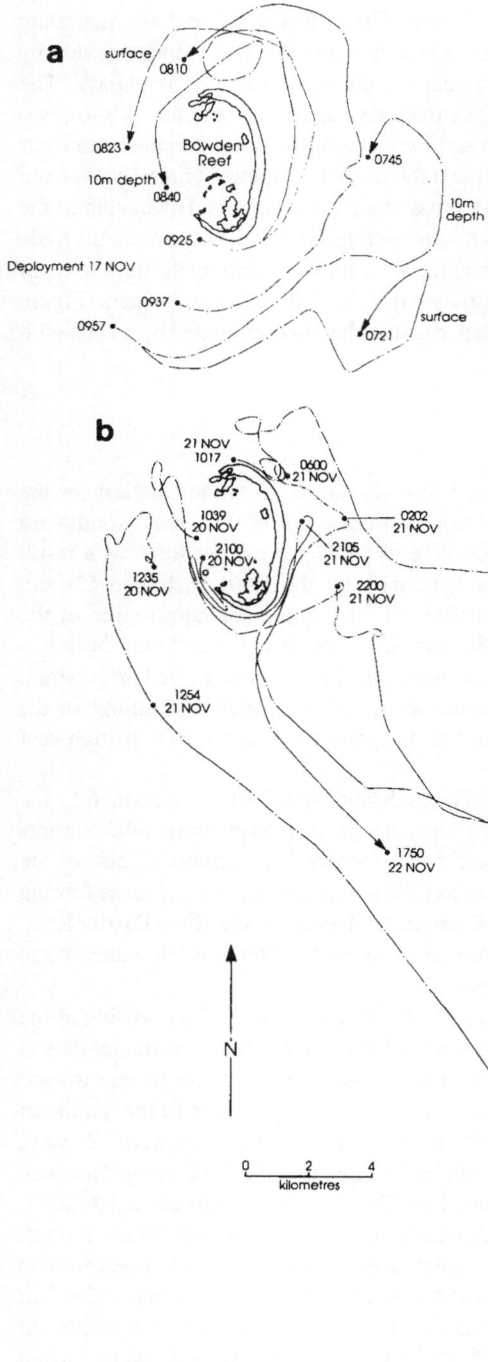

FIGURE 6.6. Trajectories of radar-traced drogues with crosses 2 × 2 m at 10-m depth and near the surface on (a) November 17 and 18, 1986, and (b) 20 to 22 November, 1986. (From Wolanski et al., 1989. Reproduced with the kind permission of Gauthier-Villars.)

around Bowden Reef at tidal period with little evidence of a mean drift. A few days later, for smaller tides on November 20 to 22, 1986, the same drogues (Figure 6.6b) were rapidly flushed from Bowden Reef and moved southeastward with the East Australian Current.

7 High-Frequency Waves

The most apparent high-frequency waves are the wind-driven waves and swell. However, there are also waves of longer period which may be due to a number of reasons, such as internal seiches, trapped waves in embayments or in lagoons, and waves due to the interaction of complex currents with a matrix of reefs.

7.1 SEICHING

Longer-period high-frequency waves can have periods on an order of hours or minutes. They are generally identified from time series of currents and sea levels by removing the tidal component. The resulting time series is then low-pass filtered to remove low-frequency fluctuations. The best evidence of such motions is in Torres Strait, where observed and predicted (by tidal harmonic) sea levels at times show significant differences, up to 0.8 m (Figure 7.1). These motions represent a significant hindrance to the navigation of large ships. Field observations of the high-frequency currents associated with these waves (Wolanski et al., 1988b) suggest that in Torres Strait, these motions reflect the passage of two-dimensional waves, but the origin of these waves is still unclear.

Current-metre data from moorings over reef flats occasionally show high-frequency motions, up to 0.1 m s^{-1}, of a period from 2 to 16 min (Wilson, 1985; Wolanski and King, 1990). The origin of these waves is unknown; it is believed that they are oscillations trapped within coral reef lagoons (Buchwald and Miles, 1981), although other explanations are possible, such as wave groups (Wilson, 1985) or resonant waves trapped around the reef slope (Massel, 1989).

Continuous tide charts of sea level at Cooktown (15.5°S), Cairns (16.9°S), and Townsville (19.3°S) in the 1980 southeast trade wind season showed two events, each lasting about 3 d, when small (<5 cm) but short-lived (20- to 60-min duration) fluctuations occurred at all three sites (Wolanski and Bennett, 1983). These fluctuations have a longer period than those expected from resonance in these harbors and surrounding embayments. Since they occurred in all three ports together, they were probably excited by a common, but unknown, external disturbance.

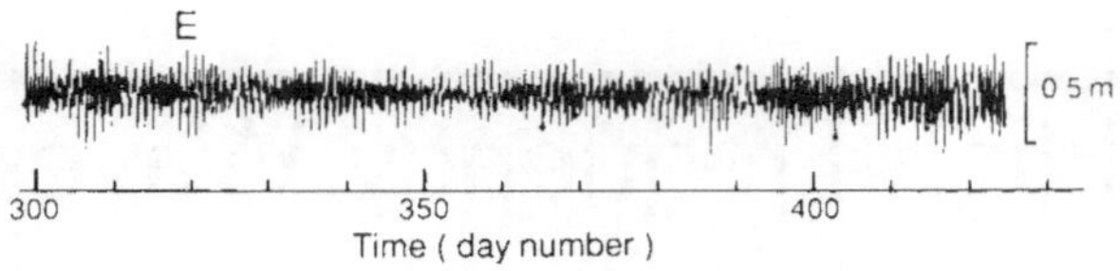

FIGURE 7.1. Time series plot of the high-frequency residual sea level at site E in Torres Strait. Location map in Figure 1.2d. Time is in day no. in 1986–1987.

7.2 FORCING

The three dominant wave-driving mechanisms are the forcing from the Coral Sea, the local wind, and tropical cyclones.

7.2.1 Forcing from the Coral Sea

By far the most obvious high-frequency waves are the wind-driven waves and swell. Wolanski (1986b) reported observations of these waves offshore from the Great Barrier Reef from a pressure gauge deployed for 3 months in the southeast trade wind season of 1980 at 17 m below the surface in a total water depth of 100 m, at the shelf break near Myrmidon Reef (18.3°S; Figure 7.2a). The 10% significant wave height fluctuated in time, peaking at 4.19 m. The relationship between significant wave height and wind depended on both the wind direction and the wind period, illustrated in Figure 7.2b. The coherence was significant only for a wind of period >4 d, a finding implying that nonequilibrium conditions prevailed between wind and sea state at periods <4 d. Figure 7.2b also shows that the coherence between wind and waves was highest and most significant for southeasterly wind and lowest for northeasterly wind. The high coherence between significant wave height and the southeasterly wind can be attributed to the practically infinite fetch of open waters in the Coral Sea from that direction and the spatial uniformity of the southeast trade winds (Figure 2.3). A low coherence existed between wave height and the wind from the northeast, and this may be attributed to the wave shadow caused by Flinders Reef located about 100 km northeast of the measurement site.

Following a shift in wind direction, the wave spectrum responds simultaneously at all frequencies, i.e., long and short waves arrive simultaneously at the mooring site. This observation implies that the mooring site is in a wave generation area. The waves are thus driven by the local wind over the Coral Sea, and do not originate from storms in far away seas, contrary to, for instance, the swell in Southern Australia originating from the Southern Ocean (Hinwood et al., 1982) and the swell in Hawaii originating from the Aleutian Islands area (Snodgrass et al., 1966).

Autospectra of the sea level (Figure 7.3) show two broad peaks at about 7 and 10 s for light winds (line a). For increasing wind, these two peaks coalesce (line b). For strong winds, the 10-s waves begin to dominate (line d). For falling winds, the 10-s waves begin to disappear first (line c). Lines b to d coalesce at periods

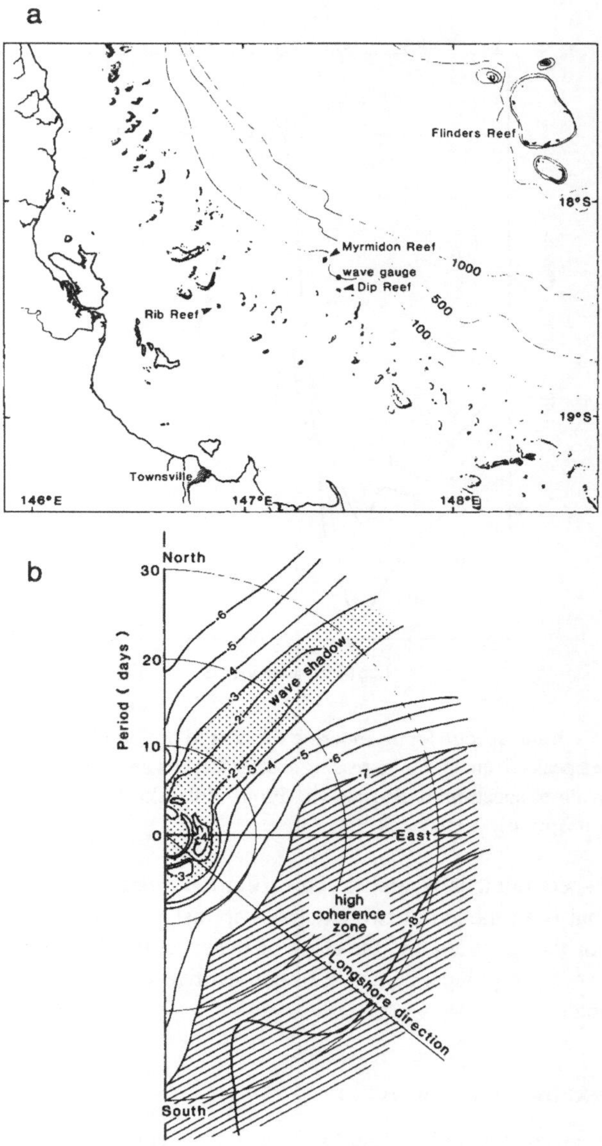

FIGURE 7.2. (a) Wave gage location and (b) plot showing the distribution for various wind directions and periods of the coherence squared (c_o^2) between the 10% significant wave height and the wind component along that direction. (Adapted from Wolanski, 1986b.)

<5 s, implying that this part of the spectrum is saturated. The slope of the spectrum for a fully developed sea (line d) is about –5, as in classical wave theory. This spectrum can be approximated by classical wave models, e.g., is well reproduced by the theoretical Wallops spectrum (Huang et al., 1981), suggesting that a

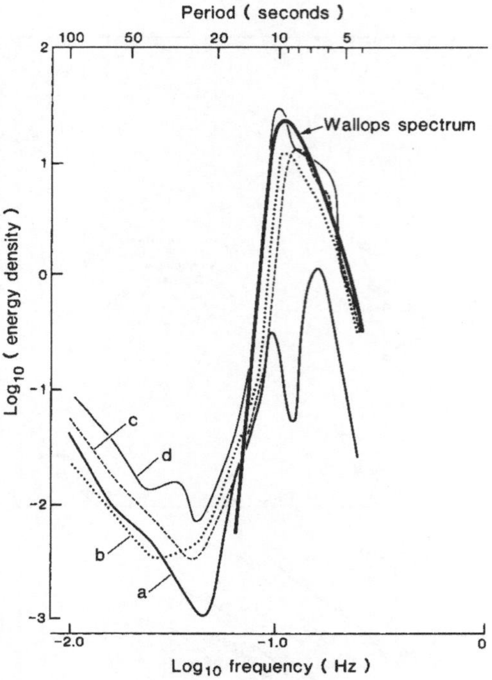

FIGURE 7.3. Wave autospectra for the mooring site near Myrmidon Reef showing (a) separate swell and wind-sea peaks (b and d) for increasing southeasterly wind and (c) for decreasing wind. The theoretical Wallops spectrum is also shown. (From Wolanski, 1986b. Reproduced with the kind permission of Springer-Verlag.)

universal spectrum locally applies. This spectrum is not a curve-fitting, empirical exercise but is a function of two external parameters, namely, the frequency for the peak of the spectrum and the slope of the spectrum. The data also show the presence of wave groups, i.e., a set of consecutive waves (typically five) larger than normal.

7.2.2 Forcing by the Local Wind

The wave shadow effect found at Myrmidon Reef (Figure 7.2b) implies that reefs significantly lower wave height for long distances in their lee. The Great Barrier Reef thus inhibits wave penetration from the Coral Sea. Waves inshore from the Great Barrier Reef matrix are thus mainly locally wind-driven waves generated in the narrow fetch of open water between the coast and the reef matrix. There is strong evidence for this in the data from wave rider buoys in shallow coastal waters. For instance, the wave spectrum in shallow coastal waters near Cairns (16.9°S) show a peak at a 4-s period, due to locally wind-driven waves, and a much smaller peak, an order of magnitude smaller, at an 8- to 9-s period which may be associated with the leakage of waves through the Great Barrier Reef matrix (Murray and Ford, 1983).

7.2.3 Forcing by Tropical Cyclones

Tropical cyclones are major wave-driving mechanisms. A predictive numerical model of tropical cyclone wind waves has been proposed by Young and Sobey (1980, 1981), and Young (1988). The model is designed for open waters with no coral reefs. The model calculates along ray paths the directional wave energy density spectrum E at a number of grid points, from a numerical solution of the wave energy conservation equation. At each grid point, E represents the superposition of free linear wave components of all frequencies and from all directions. The model calculates the rate of change of E relative to a wave group moving along wave orthogonals at the group velocity. The principal energy source or sink terms incorporated in the model through semiempirical transfer functions are atmospheric turbulent fluctuations, the instability in coupling between waves and the mean atmospheric boundary layer flow, turbulent bottom friction, and wave breaking or white capping. The model also incorporates the energy transfer within the wave field through weakly nonlinear wave-wave interactions. As an example, the predicted distribution of significant wave height in a tropical cyclone, with a central pressure of 950 mb and a forward speed of 30 km h^{-1} in open water (no topographic constraints), is shown in Figure 7.4. The vectors indicate the mean direction of the propagation of the waves.

These results have important implications for the impact of tropical cyclones on coral reefs (Kjerfve and Dinnel, 1983). The extreme wave height in tropical cyclones may cause much mechanical damage to coral reefs, but the return period

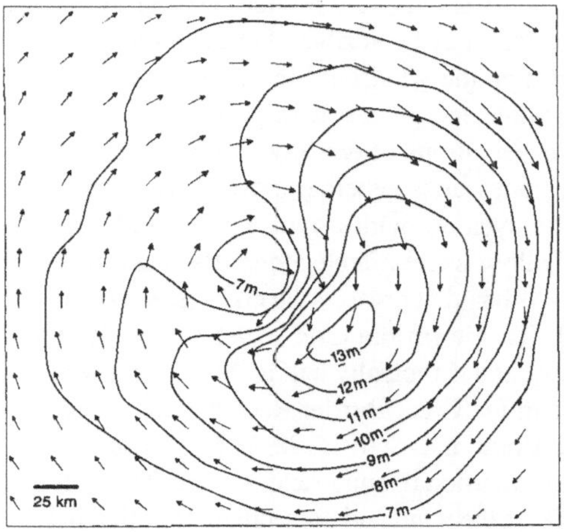

FIGURE 7.4. Distribution of the predicted significant wave height and direction of wave propagation in a tropical cyclone. The vectors indicate the mean direction of propagation of the waves. (Adapted from Young and Sobey, 1980.)

may be large. However, this damage is not necessarily related to the wind speed and it is also very sensitive to both the shelter provided by one reef to other reefs and to wave direction (Yonge, 1940; Done, 1992). Both of these parameters are extremely difficult to estimate reliably from numerical models.

Land and reef boundaries are accepted by the model as points with zero wave energy, i.e., the model assumes that the reefs dissipate all the incident wave energy that is not channelized in the reef-free waters. The small-scale topographic complexity, as well as reflection and refraction effects around coral reefs, invalidate some of the assumptions of the model when it is applied to the Great Barrier Reef.

Gourley and McMonagle (1989) used the model to hindcast waves in the open waters of the Capricorn Channel following cyclone David in January 1976. A significant wave height of about 4 m with a period of about 7 s was measured by a wave-rider buoy in coastal waters. The predictions overestimate by 30% the wave height observations. The discrepancy may be due to unresolved problems in parameterizing the sheltering effect of reefs, and to open boundary conditions in the far field. Young and Hardy (1993) repeated a similar exercise for tropical cyclone Aivu that crossed the coast in April 1989. They modeled the wave climate in the Coral Sea using a mesh size of 50 km. As the cyclone approached the Great Barrier Reef, smaller mesh sizes of 25 km and finally 2.5 km were used. Data from wave-rider buoys were available at two sites, one at Leopard Reef (19.1°S) located near the shelf break, 75 km southeast of the track of the cyclone, and one in the Great Barrier Reef matrix near John Brewer Reef (18.5°S), located 80 km northwest of the cyclone's track. At the Leopard Reef mooring in the open ocean, the significant wave height was overpredicted by about 25%, while the peak wave period, varying between 10 and 13 s was adequately reproduced by the model. The significant wave height at John Brewer Reef, where the wind was from the coast and the fetch was thus very limited, was also overpredicted by 20 to 30%. Some of these errors may be attributed to the model being extremely sensitive to errors in hindcasting the wind field in the tropical cyclone. The model suggests that some long-wave energy is able to penetrate through reef passages, in agreement with the observations of Murray and Ford (1983).

Tropical cyclones also generate storm surges. A surge is a trapped wave moving with the cyclone, with a typical amplitude of <0.5 m on the shelf but amplified by shallow water effects along the coast to several metres, e.g., 3.6 m during cyclone Althea that crossed the coast near Townsville (19.2°S) on Christmas Eve, 1971. The storm duration is determined by the speed of the cyclone crossing the coast and is typically half an hour to a few hours. A numerical model of storm surges in the Great Barrier Reef region has been proposed by Sobey et al. (1982) and Bode and Stark (1983). They used a finite-difference, depth-averaged, two-dimensional numerical model. The forcing was provided by a parameterization of the wind and pressure distribution in the moving cyclone. The reefs were modeled as weirs. The mesh size was 5 nmi ($\simeq$ 9 km), so no details of the interreefal circulation emerged. The model appears reliable in its predictions of storm surge height but seems to overpredict, by 50%, the observed peak longshore coastal currents (Wolanski and Ridd, 1990). This may be due to an underestimate within the models of the radiation stress resulting from wave-current interaction (Grant and Marsden, 1979; Grant et al., 1984). This effect increases the bulk friction parameter for the storm surge currents.

7.3 WAVE BREAKING BY CORAL REEFS

The mechanisms by which waves are attenuated by, and filter through, the Great Barrier Reef are still poorly understood. One-dimensional effects are simplest to understand: wave shoal and break on the reef edge and then propagate, with new frequencies and amplitude, on the reef flat. Two-dimensional effects occur when waves are refracted and reflected around coral reefs. These processes are still poorly understood.

One-dimensional effects (i.e., wave breaking on a reef of practically infinite length) have been identified as a mechanism generating a net circulation over the reef flat at Bikini Atoll (Munk and Sargent, 1954). Roberts et al. (1975) estimated that up to 95% of wave energy is dissipated by wave breaking at Grand Cayman Island. Field observations in the Great Barrier Reef were initiated by Pickard (1986), who, in the absence of wave data, subtracted the tidal and wind-driven components from the current-metre data to calculate the wave-induced currents over the windward reef flat at Davies Reef (18.7°S). He estimated that wave overtopping may contribute 40% of the flow over the windward reef flat at Davies Reef for winds of 7 m s^{-1}. Hardy et al. (1990) deployed water-level recorders, current metres, and wave recorders on a cross-reef slope transect across the exposed reef slope and reef flat of John Brewer Reef (18.6°S). Incident waves were measured by a wave-rider buoy located about 500 m seaward from the reef front in a depth of 50 m. Four wave poles were deployed for a total of about 45 d on the reef flat, from 27 to 50 m from the reef edge. The first wave pole was in the breaker zone. Finally, another wave-rider buoy was deployed in the semienclosed lagoon of the reef. The significant wave height offshore was found to be independent of tidal height. A significant reduction of wave height from offshore to the reef flat was observed. This reduction was a function of the water depth over the reef flat, the significant wave height over the reef being higher at high tide than at low tide. Not all the wave energy is transformed in the first 27 m of the reef flat as the significant wave height varied across the reef flat, the value of the ratio of significant wave height to water depth varying from 0.7 at 27 m from the reef flat to 0.4 at 50 m from the reef flat. This ratio was not a constant at a given site; at 50 m from the reef flat, the ratio decreased with increasing water depth from a value of $\approx$0.4 at low tide when the water depth was <0.5 m to a value of about 0.24 at high tide when the water depth was >2.5 m.

This study clearly demonstrated that a significant fraction of the incoming wave energy is dissipated by a reef, in agreement with earlier field and laboratory studies of wave breaking and transformation on submerged breakwaters, steep shallow shelves, and reefs elsewhere in Australia, Japan and the U.S. (e.g., Gerritsen, 1981; Horikawa and Kuo, 1967; Lee and Black, 1979; Nelson and Lesleighter, 1985; Roberts et al., 1975, 1977; Roberts, 1981; Young, 1989).

Wave refraction-diffraction models have been proposed for many years for studying the wave transformation on a slowly varying topography such as a beach. Such models are not applicable to coral reefs because the mild slope assumption of the models is not consistent with the steep slope of the reefs. Massel (1992) elegantly removed the mild slope restriction. He assumed that wave breaking and bottom friction are the dominant effects responsible for energy dissipation. He modeled a one-dimensional reef, with steep slopes approaching 1:1, and a wide fringing reef flat (Figure 7.5a). Nonlinear wave-wave interactions between different spectral components are neglected, as well as the possible generation of higher harmonics waves on the reef. The waves have a single dominant frequency but a random wave height. The latter is calculated from the refraction-diffraction equation, which requires the specification of the average rate of energy dissipation due to wave breaking and bottom friction. It is assumed that these two effects can be computed separately and added linearly. These two dissipation rates are parameterized from semiempirical laboratory data. The dissipation of energy by wave breaking is parameterized by multiplying the rate at which energy is dissipated by a single wave-breaking event (a bore) by the probability of the wave breaking at each height. The dissipation of energy by friction in the wave boundary layer is calculated by the friction law of Gerritsen (1981). Finally, the radiation stress due to shoaling, refraction, diffraction, and dissipation introduces a sea-level rise that is calculated from the momentum equation. The variation with distance x of the significant wave height H_{rms} and mean sea level setup η is shown in Figure 7.5b for an incident wave with significant wave height $H_{rms} = 4$ m with period $Tp = 8$ s. This figure also shows

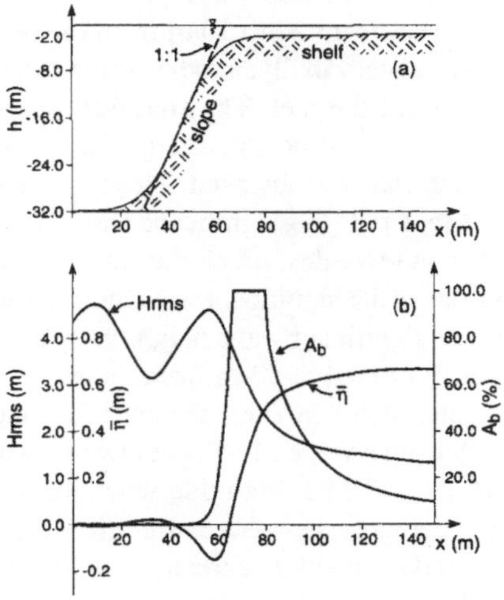

FIGURE 7.5. (a) Model geometry and (b) predicted cross-reef distribution of average wave height (H_{rms}), mean sea level (η), and percentage of wave breaking (A_b) for an incidental H_{rms} = 4 m and a wave period of 8 s. (Adapted from Massel, 1992.)

the distribution of A_b, which is the percentage of breaking waves. The oscillations of H_{rms} at 10 to 60 m from the reef edge are due to the interference of incident and reflected waves. All waves break on the reef edge. The waves reform downstream so that less than 2% of the waves are breaking at 150 m downstream from the reef edge where, however, the wave amplitude is very small. The wave setup is considerable, on the order of 0.6 m or about 17% of the incident wave height, and occurs a few metres downstream of the reef edge.

A most important result in that most (in this example about 88%) of the wave energy is dissipated. Most of that dissipation can be attributed to wave breaking, not bottom friction, suggesting that the simplifying assumptions behind the model are justified in hindsight and that the model may find general applicability.

7.4 WAVE SETUP AND CIRCULATION

One important implication of wave breaking on the reef crest is the generation of a sea-level setup (Figure 7.5b). This sea-level setup generates a circulation over the reef flat, a simple momentum balance prevailing between the sea-level gradient, the direct wind stress, and the bottom stress. Prager (1991) and Wolanski et al. (1993) modeled this effect by adding, in the cross-reef horizontal momentum equations of a depth-averaged, two-dimensional model, and at predetermined points where waves break, a positive stress representing the transfer of momentum from wave breaking from the radiation stress

$$S_{xx} = 3/16\ \rho\ g\ H_{rms}^2 (\mathrm{Nm}^{-1}) \tag{7.1}$$

where ρ is the water density, g the acceleration due to gravity, and H_{rms} the average wave height. The method was applied to Great Pond Bay, St. Croix, Virgin Islands, and the fringing reef of Moorea Island, French Polynesia. The results suggest that a 4 m wave breaking over the reef crest locally applies the same stress as a 50 m s^{-1} wind. Field data are sketchy, allowing only qualitative, but encouraging, comparisons between observed and computed currents.

This effect has not received much attention in studies of the water circulation over reef flats in the Great Barrier Reef. Wilson (1985) also proposed that a forcing term be included to represent momentum transfer by wave breaking on the reef flat. By the use of a semiempirical term, he could reproduce qualitatively the tidal current asymmetry over the reef flat at One Tree Island (23.5°S) previously reported by Davies and West (1981).

7.5 WAVE FOCUSING BY REEF PLATFORMS

On platform reefs, the assumption of one dimensionality breaks down. Wave energy can be focused at the back of the reef platform (Figure 7.6). This energy comprises that of two sets of waves: (1) those that break on the reef crest can reform over the reef platform and travel across the reef platform, where they are

FIGURE 7.6. Photograph of wave focusing on a reef flat.

refracted, and (2) those traveling on the slope around the platform which are diffracted into the shadow behind the platform (Gourlay, 1988). The oceanic waves diffracted around the platform reef are deep-water waves and have a wavelength and phase speed independent of depth. The waves over the platform are shallow-water waves, with a celerity that is dependent on the depth h over the platform.

These wave-induced dynamics are responsible for the creation and maintenance of coral cays (islands on reef platforms; Figure 7.7). The size, shape, and orientation of the reef platform determines whether a cay will form and is stable (Hopley, 1982: Gourlay, 1988). The cays are generally more stable when they are oval or elliptical, with their long axis oriented approximately along the dominant incident wave direction, since this configuration locates the wave focus point

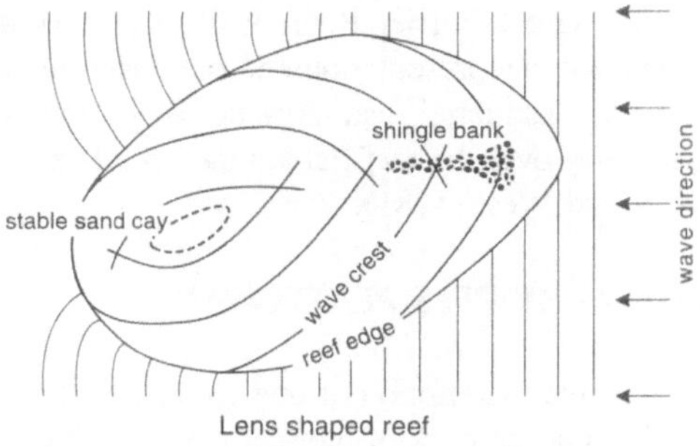

FIGURE 7.7. Wave refraction on coral reefs. (Adapted from Gourlay, 1988.)

quite stably, more or less independent of small changes in wave direction. On such a reef platform, the formation of a long sand bank or windward shingle cay may also be associated with the meeting of waves created by waves on either side. Cays on circular reefs are less stable, since the wave focus point can readily move with changes in wave direction.

Quantitative models for such processes are still unavailable. Yet, predictive, reliable models of the movement of water and sediment on coral reef platforms are urgently needed to assess the fate of sewage discharge and the transport of natural sediment and dredge spoil because ill-planned tourism developments have proceeded on such cays, as at Heron Island (23.5°S) and Green Island (16.6°S), with significant environmental degradation.

7.6 BIOLOGICAL WAVE DAMPING

There is evidence that corals respond to water disturbance by exuding and sloughing off mucus, and this release may increase with increasing wave activity (Benson and Muscatine, 1974; Johannes, 1967). A sufficient quantity of the surfactant from the coral mucus may calm the sea surface, even under moderately strong winds, by preferentially damping the capillary waves, and this may explain the observations that the wind drag coefficient can be smaller by a factor of 2.6 over a coral reef than in the open sea (Hicks et al., 1974; Deacon, 1979). This effect has received surprisingly little attention by reef oceanographers.

8 Reef-Induced Circulation

Flows around coral reefs are very complex and occur at all scales. At a scale of tens of kilometres, a dense matrix of reefs can, by increasing friction, strongly inhibit the water circulation, such as happens in Torres Strait. At a scale of kilometres, flows around reefs generate complex eddies and jets, which generate patchiness and streakiness. These motions are strongly three-dimensional and lead to tidal pumping and localized upwelling events. At smaller scales, flows within reef-lagoon systems generate complex recirculating zones and trapping phenomena. At a scale of metres, flows around the complex topography of a sloping reef surface also generate complex boundary mixing phenomena and small-scale upwelling and shear zones which are biologically important. The various scales are discussed, starting with the flows around an individual coral reef.

8.1 REYNOLDS NUMBER ANALOGY

Steady, barotropic flows around an obstacle (a plate) in laboratory experiments (Figure 8.1) are governed by the Reynolds number

$$R_e = U\,W/\nu \tag{8.1}$$

where U is the free-stream velocity, W (W = 2R) is the width of the obstacle, and ν is the kinematic viscosity of the water (Batchelor, 1967; Gerrard, 1978). The depth H is not considered because the experiments are carried out for H >> W, so that the resulting mean motions are two-dimensional (depth independent) while the turbulence is three-dimensional. For $R_e < 1$, there is no flow separation and no eddy forms. For $R_e > 1$, a steady wake is formed comprised of two standing eddies and a zone of return flow (Figure 8.1a); the velocity within the eddy is very small, typically 1% of the free-stream velocity (Keller and Niewstadt, 1973). For $R_e > 20$, flow instabilities (meanders) develop on the interface between wake waters and the surrounding waters. As R_e is further increased, these instabilities become meanders far downstream (Figure 8.1b). For even higher values of R_e , the meanders can become unstable (Figure 8.1c). Finally, Karman vortices are shed. The frequency n at which eddies are shed is related to the forcing frequency U/W, so that the Strouhal number

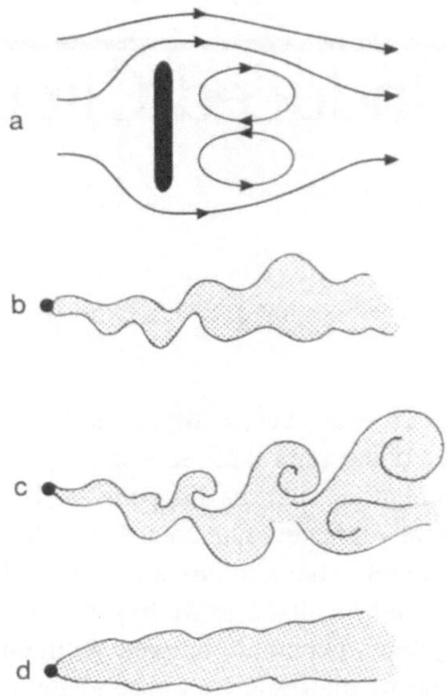

FIGURE 8.1. Plan view of the water circulation around an obstacle.

$$S_t = W\,n/U = 0.21 \text{ for } R_e > 300 \tag{8.2}$$

The eddies move at a speed of about 0.85 U and have a spacing of about 5W, but this distance increases downstream.

For $R_e > 2500$, the wake is fully turbulent far downstream with no organized structure (Figure 8.1d).

Baines and Davies (1980) undertook laboratory experiments to extend this model to the case of rotating flows. The relevant parameters turn out to be the Rossby number ($R_o = U/2\,f\,R$) and the Ekman number ($E_k = \nu/2\,f\,R^2$), where f is the Coriolis parameter. The Reynolds number is the ratio of Rossby to Ekman numbers, so their theory is a generalization of the above one. Boyer and Davies (1982) extended the model to consider latitudinal variations of the Coriolis parameter (β-plane approximation). These earth-rotation effects appear to be unimportant for flows around coral reefs on the continental shelf of the Great Barrier Reef because the rotation rate ω of water in the eddy is much larger than f (Tomczak, 1988).

8.2 ISLAND WAKES

Aerial and satellite pictures of flows around reefs and islands in the Great Barrier Reef show island wakes (e.g., Figure 8.2a, b) that have a shape similar to that in laboratory experiments for Re = 5 to 10. For typical values, $U = 0.6$ m s^{-1} and $W = 2000$ m, Equation 8.1 yields $\nu = 1.2$ to 2.4×10^2 m^2 s^{-1}, a value much greater than the expected value (0.1 to 1.0 $m^2 s^{-1}$ and 1 to 10 $m^2 s^{-1}$) of, respectively, the vertical and horizontal eddy viscosity coefficients in shallow coastal waters for eddy motions at such scales (Okubo, 1974; Fischer et al., 1979). In other words, for an eddy viscosity of the order of 3 $m^2 s^{-1}$, the Reynolds number is the order of 400, which in laboratory experiments implies a long turbulent wake, while the field observations show an organized eddy structure. Hence, the Reynolds number is not a suitable parameter for describing flows around coral reefs in the Great Barrier Reef. The difference between field conditions and the laboratory experiments is the aspect ratio, i.e., the ratio W/H. In laboratory experiments, W/H << 1 and bottom friction is not important; for islands and coral reefs in the Great Barrier Reef, W/H >> 1 and bottom friction may become important.

Little was known of the dynamics of flows around coral reefs and islands until a study was carried out at Rattray Island (20°S; Figure 8.3) by Wolanski et al. (1984a). The island is 1.5 km long and 300 m wide and lies in well-mixed coastal waters 20 to 30 m deep. Its long axis is oriented at about 60° to the direction of the dominant semidiurnal flood tidal current. Flood tidal currents are southeastward longshore and generate an island wake in the lee of the island (Figure 8.2b). Twenty-six current metres were deployed along four transects (Figure 8.3); the transects crossed the island wake at flood tidal currents, and the current metres were located upstream of the island during ebb tidal currents. The hourly synoptic distribution of the measured currents on December 4, 1982, is shown in Figure 8.4. At 8 h at the start of the flood tide, the currents were sluggish and there was no eddy in the lee of the island. An eddy started to appear at 10 h and grew in size until 12 h. When the surrounding tidal currents decreased at the turn of the tide, the eddy started to spin down but lagged behind the surrounding currents. A small eddy was still present at slack tidal current (14 h). The eddy was then advected away at the start of the ebb tide (15 h). Two hours into the ebb tide (at 16 h), the currents upstream of the island were undisturbed by the island except for the metres closest to the island.

The eddy had a small thermal signature, trapping slightly colder water than the surrounding waters (Figure 3.12b), although the waters were vertically well-mixed. It also had a strong three-dimensional component. Fine sediment was upwelled in the center of the eddy. Drogues released in the eddy moved at 0.1 to 0.2 m s^{-1} toward the separation point on the northwest side of the island, where they stagnated for several hours, indicating that surface water was advected toward the separation point, where it downwelled rapidly. There was also a marked horizontal velocity shear near the separation point.

Contrary to the laboratory experiments, eddies behind islands have a large aspect ratio (W/H >> 1) and a three-dimensional circulation with zones of strong upwelling and downwelling. Another key difference is that in the laboratory, u/U << 1, while at Rattray Island, u/U $\approx$ 1, where u is the peak velocity within the eddy and U is the undisturbed velocity. Thus, the Reynolds number analogy is not valid.

These observations suggest a three-dimensional flow model (Figure 8.5) where at steady state a balance exists between vorticity input into the eddy from the shear

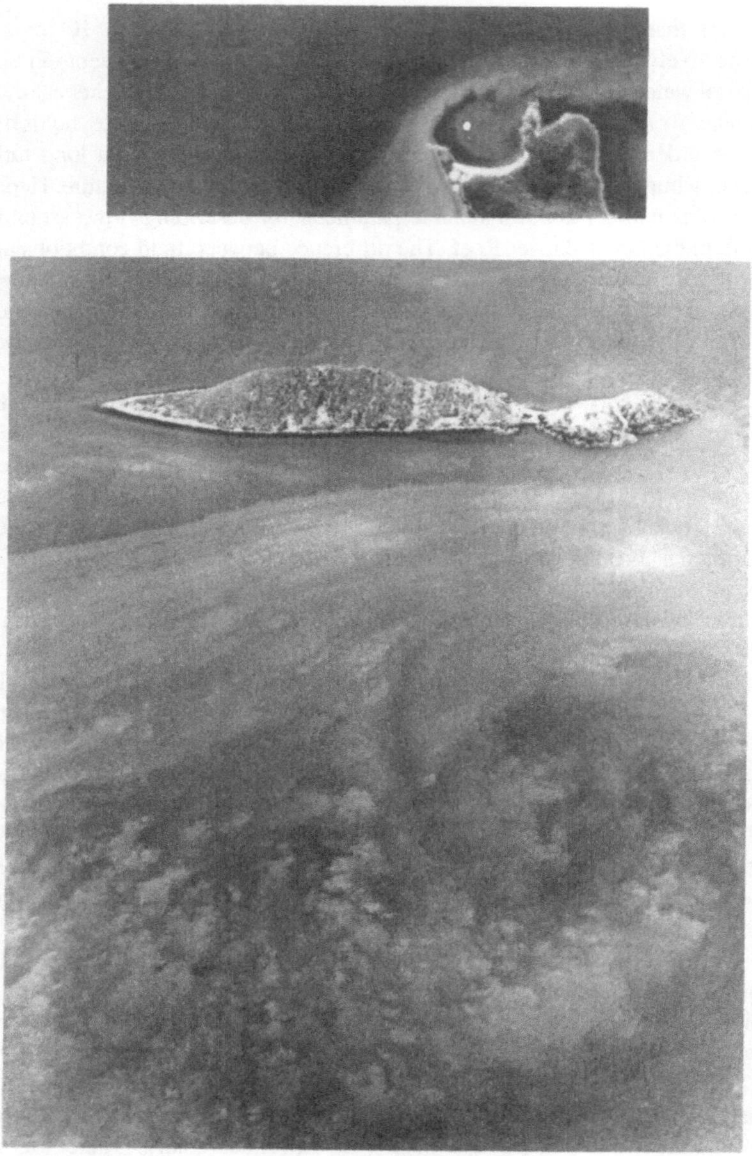

FIGURE 8.2. Aerial photographs of eddies shed by island and headlands in the Great Barrier Reef. (Figure(s) 8.2c to e from Wolanski, 1986c. Reproduced with the kind permission of Springer-Verlag.)

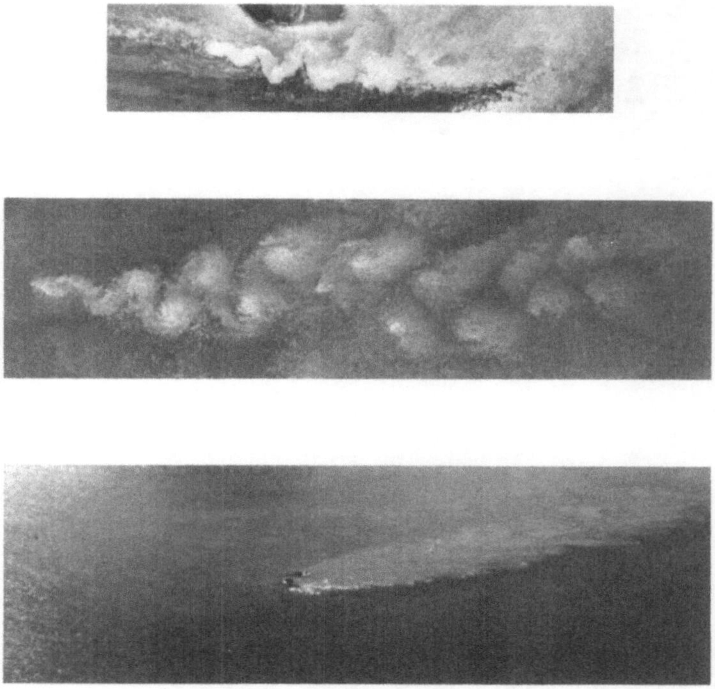

FIGURE 8.2 (continued).

layer at the separation point, and the vorticity loss by bottom friction in the eddy (Wolanski et al., 1984a). Similar to the circulation in a tea cup, the secondary circulation is characterized by upwelling at the center. This secondary circulation advects water from the eddy toward the bottom, where vorticity is lost through friction and the fluid is spun down. For model simplicity, it is assumed that there are two distinct regions in the eddy: (1) a core region where solid body rotation prevails and (2) a bottom boundary layer of thickness δ. It is also assumed that $\delta \ll H$.

Solid body rotation is assumed in the core of the eddy.

$$v = \omega r \tag{8.3}$$

where v is the azimuthal velocity, r is the radius, and ω the vorticity. The momentum balance in the bulk of the fluid, outside the bottom boundary layer, is

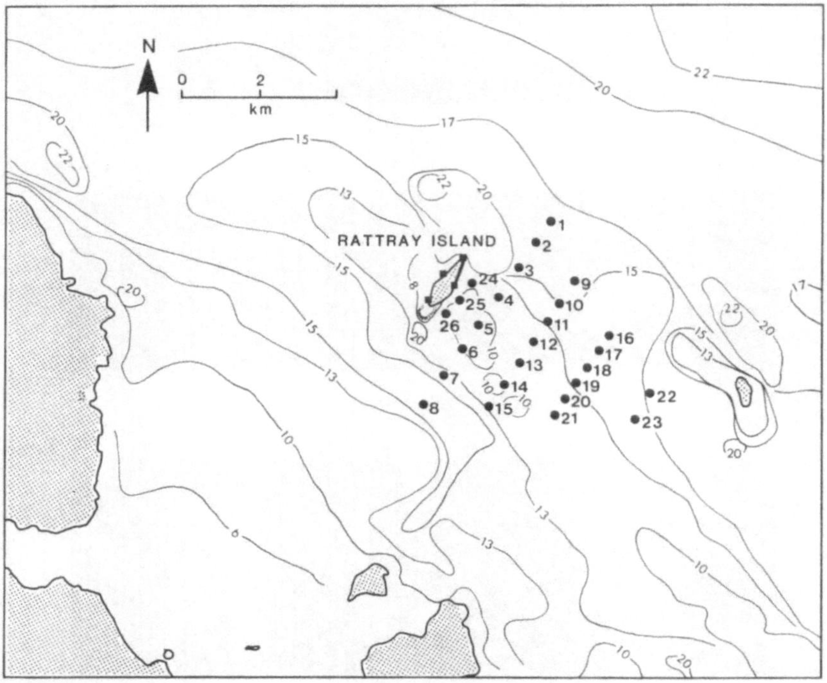

FIGURE 8.3. Rattray Island, location of current-metre moorings (.) and tide gages (squares), depths in fathoms (1 fathom = 1.8 m). (Adapted from Wolanski et al., 1984a.)

$$v^2/r = g\, d\eta/dr \tag{8.4}$$

where g is the acceleration due to gravity and η is the free surface elevation. This balance breaks down in the bottom boundary layer where the pressure term g $d\eta/dr$ still applies but where $v \to 0$ by friction. The pressure gradient generates in the boundary layer a radial component of the current, u, such that (Greenspan, 1968)

$$v^2/r = K_z\, d^2u/dz^2 = O(K_z u/\delta^2) \tag{8.5}$$

where K_z is the vertical eddy diffusion. Following Fischer et al. (1979),

$$K_z \simeq 0.067\, Hu_* \tag{8.6}$$

where u_* is the shear velocity ($\simeq 0.05$ V). Conservation of mass in the boundary layer requires that $V \simeq U$. This leads to the scaling law for the thickness of the bottom boundary layer

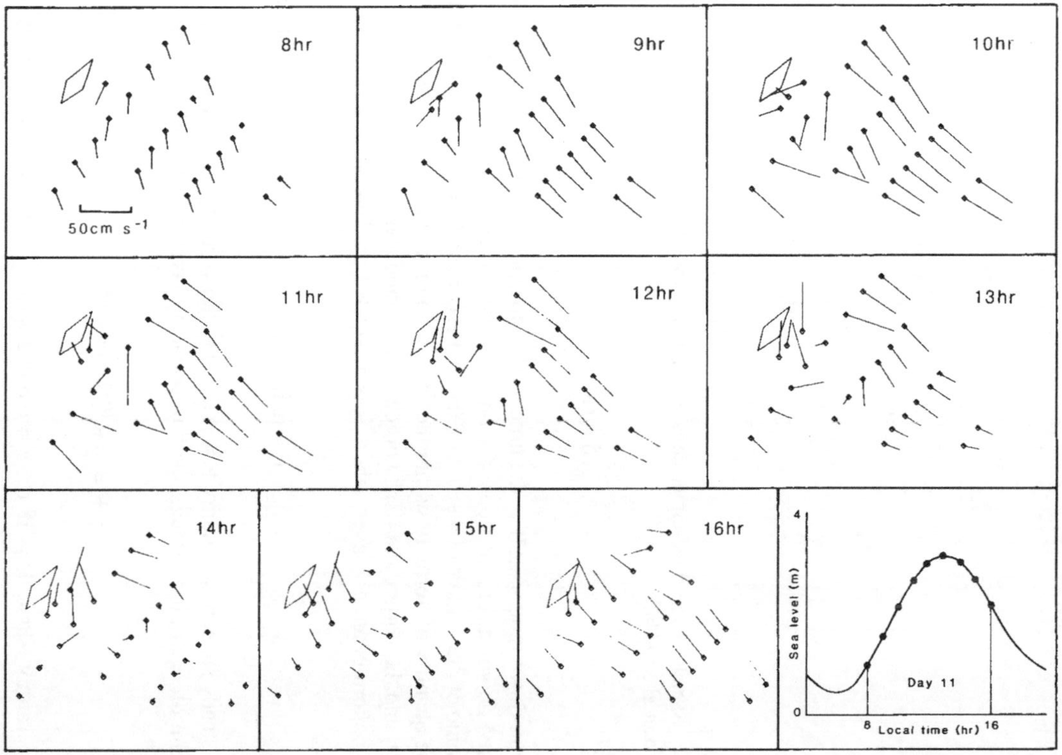

FIGURE 8.4. Hourly synoptic maps of observed currents around Rattray Island, December 4, 1982. (Adapted from Wolanski et al., 1984a.)

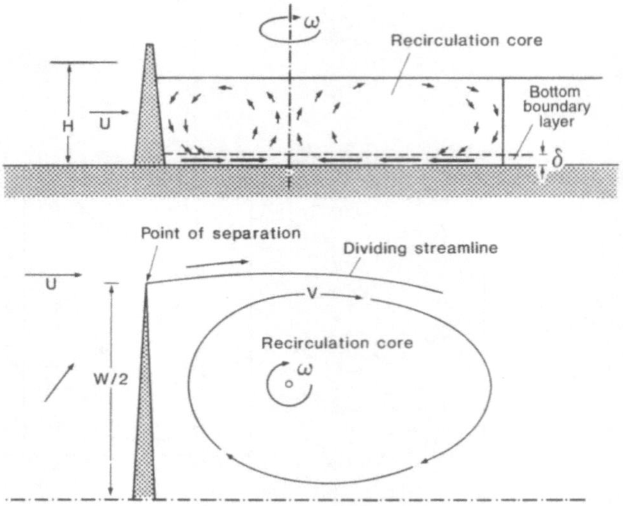

FIGURE 8.5. Sketch of the three-dimensional circulation around Rattray Island. (Adapted from Wolanski et al., 1984a.)

$$\delta = (K_z/\omega) \tag{8.7}$$

The spin-down time scale T is the time it takes for the water from the core to be advected downward to the boundary layer, where it is spun down by friction. T is the ratio of the volume of water, $O(R^2H)$, in the core and the flux of water, $O(uR\ \delta)$. The quantity of vorticity within the core region is the product of the volume $O(R^2H)$ and the vorticity ω. This vorticity is advected away from the core region to the bottom boundary layer, where it is dissipated by friction, at a rate

$$F = \omega\, R^2 H/T \tag{8.8}$$

In the steady state, this vorticity loss is equal to the vorticity entering the eddy at the point of separation through the thin boundary layer, at a rate $O(HU^2)$. Then

$$F = \omega\, R^{1/2} V/\omega^{3/4} K_z^{1/4} \tag{8.9}$$

For Rattray Island, H = 18 m, V = 0.6 m s^{-1}, $\omega = 1 \times 10^3\ s^{-1}$, and it results in R = 1043 m. This prediction compares well with the observed length of the wake (Figure 8.4). The time scale T = 0.85 h, which suggests that a quasi steady state prevails. The predicted internal circulation reproduces qualitatively well the observed upwelling within the eddy center and the downwelling along the island slopes. It also explains the accumulation of floating material as a foam line along the outer edges of the eddies in shallow, well-mixed coastal waters (Figure 3.16).

The scaling suggests $\delta = 16$ m, implying that the boundary layer is not very thin compared to the depth and that a more sophisticated model is needed for detailed predictions. This scaling only applies for flows with a transverse shear scale <1000 m (Geyer, 1993).

The eddy behind Rattray Island dissipates about half of the incoming kinetic energy inflow of the water facing the island upstream. This finding implies that a reef matrix with numerous scattered reefs dissipates, through the island wake effect, a significant fraction of the tidal energy.

Note in Equation 8.9 that R is independent of W, so that the eddy size is not necessarily equal to the width of the island or headland. The ratio

$$P = R/W = UW/K_z(H/W)^2 \tag{8.10}$$

is the island wake parameter (Wolanski et al., 1984a; Pattiaratchi et al., 1987). This scaling law is supported by laboratory experiments conducted by von Riegels (1938) in a Hele-Shaw cell, a flume where the flow is constrained between two closely spaced plates so that $H/W \ll 1$, and therefore friction effects dominate. For $P < 1$, the eddy is small compared to the obstacle width (e.g., Figure 8.2a). For $P = 1$, the eddy has grown to its maximum dimension, on the order of the obstacle width ($R = W$), but the flow is stable because the vorticity shed at the separation point can be dissipated in the eddy. This is the case at Rattray Island (Figure 8.2b). For $P > 1$, the analytical model suggests $R > W$. This is unrealistic and implies that the vorticity shed at the separation point cannot all be dissipated by the eddy. This vorticity is advected downstream, where the flow instabilities develop for increasing values of P. For P = 1 to 3, meanders develop downstream (Figure 8.2c). For P = 3 to 15, the meanders develop instabilities and roll (Figure 8.2d). For $P > 10$, the wake is fully turbulent far downstream (Figure 8.2e).

The model does not describe how the vorticity is injected in the eddy, nor the time it takes for eddy spin-up. Vorticity is, of course, related to flow separation. The importance of flow separation was demonstrated later by Geyer and Signell (1990), who used a ship-mounted Acoustic Doppler Current Meter to map the tidal eddy behind a coastal headland at Gay Head, Massachusetts. The flow is tidal and the eddy lags the tidal currents by up to 3 h. The vorticity distribution (Figure 8.6), calculated from the velocity data, shows that the flow separation at the tip of the island generates a very strong shear in open waters downstream of the separation point, as the velocity changes by almost 0.4 m s^{-1} in a horizontal distance of about 160 m. The magnitude of this vorticity within the shear zone exceeds 20×10^{-4} s^{-1}, five times larger than the peak vorticity within the eddy. The position of this patch of vorticity indicates that the vorticity is produced at the headland and advected into the interior of the flow. The vorticity is not constant in the eddy, implying that the eddy waters are not simply in solid body rotation as assumed in Equation 8.3, but that both rotational (solid body rotation) and irrotational motions are present.

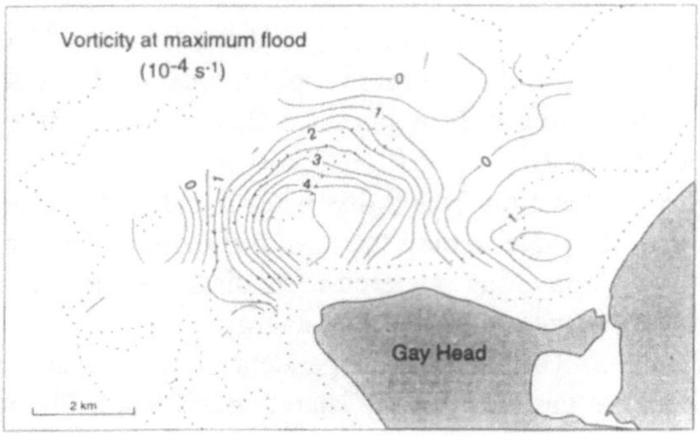

FIGURE 8.6. Distribution of the vorticity at maximum flood around Gay Head. (From Geyer and Signell, 1990. Reproduced with the kind permission of the American Geophysical Union.)

The processes leading to eddy formation behind a headland have been modeled by Singell and Geyer (1990) using a two-dimensional, nonlinear, depth-averaged numerical model and a very small mesh size. They used curvilinear coordinates with a variable mesh size as small as 53 m near the tip of the model headland, which was about 3 km wide, so that free-shear layers (vortex sheets) downstream of the headland were explicitly calculated. Predicted flow and vorticity field (Figure 8.7) suggest that the eddy starts to form when vorticity is introduced downstream of the headland by the vortex sheet rolling into itself. Middleton et al. (1993) found that an eddy does not form in the lee of a headland in the presence of a rocky promontory about 200 m offshore from the headland. They attributed this effect to the rocky promontory enhancing the turbulence and moving the separation point further downstream. Another explanation is that a jet develops in the 200-m gap between the headland and the promontory, preventing the formation of a steady eddy (see Section 8.6).

Little is known about how to parameterize the effects of these subgrid scale shear layers in coarse scale models, i.e., models where an island is represented by no more than five to 10 grid points, so that free-shear layers are not explicitly calculated by the model. Falconer et al. (1986a, b) used a two-dimensional (depth-averaged) nonlinear model to study the circulation around Rattray Island. Their mesh size was 200 m, so the island width was represented by about seven grid points. They recognized that the Reynolds stress in the momentum equations has two components: (1) bed-generated turbulence and (2) free-shear-generated turbulence. The influence of the latter dominates near the separation point. To parameterize the effects of the latter, the width of the free-shear layer (the vortex sheet) and the velocity profile within this layer were assumed to follow turbulent jet theory (Figure 8.8). In particular, the velocity U was represented by

$$U = U_1(1 + \mathrm{erf}\sqrt{R\gamma/2}) \qquad (8.11)$$

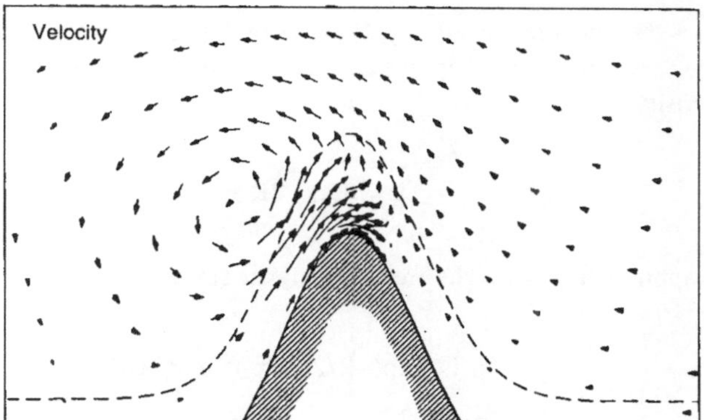

FIGURE 8.7. Predicted distribution of the velocity and vorticity around a headland. (From Signell and Geyer, 1991. Reproduced with the kind permission of the American Geophysical Union.)

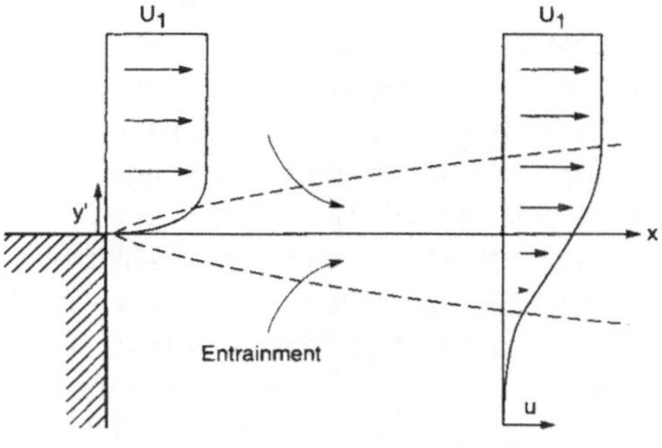

FIGURE 8.8. Assumed velocity profile downstream of a headland.

where U_1 is the free-stream velocity, R (= about 288) is an experimental constant, and $\gamma = y'/x'$, where y' and x' are the coordinate axes. This corresponds to a mean eddy viscosity

$$\nu = U_1 x'/2R \tag{8.12}$$

which is applied at $y' = 0$. The free-shear layer stress is

$$\tau_{xy} = \rho\, U_1^2 \exp(-R/2(y'/x')^2/2\sqrt{(2UR)} \tag{8.13}$$

This effect is in addition to the bed-generated component of the shear stress

$$\tau_{xy} = \rho(0.16\, u_*/C)\, \partial U/\partial y \tag{8.14}$$

where u_* is the shear velocity ($u_* = \sqrt{(gq/HC)}$, where H is the depth, q = UH, C is the Chezy friction coefficient, and $\partial U/\partial y$ is calculated by a finite difference representation from the local values of the numerically predicted velocities. The two Reynolds stresses (Equations 8.13 and 8.14) are added along $y' = 0$ for $x' < 2C^2H/\pi g$. Beyond that point, only Equation 8.14 applies.

This method appears to be successful, since predicted and observed currents at the 26 current-metre moorings at Rattray Island compare favorably with each other (Figure 8.9). The synoptic distribution of the predicted currents shows the presence of two eddies, a feature of the flow that could not be resolved by the array of current metres, one eddy being much larger than the other one in view of the inclination of Rattray Island into the flood tidal current. The predicted distribution of vorticity (Figure 8.10) calculated at the grid scale of 200 m (>> the observed width of the free-shear layers near the separation points) shows a large velocity shear between the waters outside and inside the eddy near the separation point. It also shows that the eddy is not in simple solid body rotation.

In unsteady tidal flows around islands of the size of Rattray Island, the parameterization of free-shear layers accounts for about 20 to 30% of the vortex strength (Falconer et al., 1986a). However, in steady flows, or for flows behind small headlands where the time scales of eddy formation are small compared with the tidal period, the parameterization of free-shear layers accounts for 50 to 60% of the eddy strength (Wolanski, 1987).

There are several unresolved problems in parameterizing subgrid scale free-shear layers (vortex sheets) in such coarse grid numerical models. It is not known, for instance, how to parameterize the folding of a vortex sheet. The location of the separation streamline must be known in advance. In practice, this can only be done in areas where the tidal currents change 180° with the tides and do not rotate. Unfortunately, on the shelf, the tidal ellipses show significant rotation and the

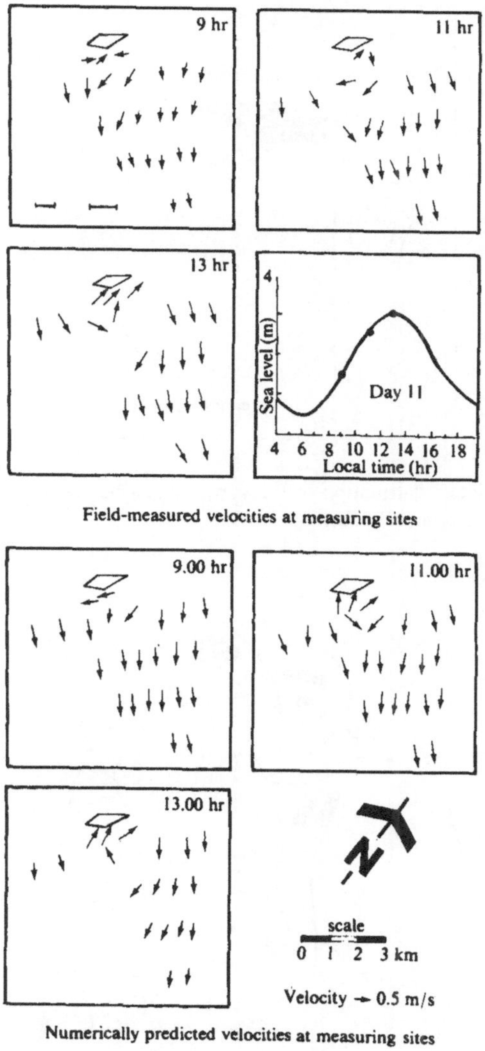

FIGURE 8.9. Comparison of observed and predicted velocity distribution at 2-h intervals at flood tide around Rattray Island. (Adapted from Falconer et al., 1986a.)

free-shear layers rotate with the tidal currents (Figure 8.11). They are readily visible, as they show up (in calm weather) as a turbulent front made visible by long lines on the water. At present, the only way to predict the location of such layers may be to use a very fine mesh numerical model (Signell and Geyer, 1990) or to rely on field observations.

Numerical models do not need a very fine mesh to generate an eddy behind a headland in unsteady tidal currents. An eddy can be produced with only about

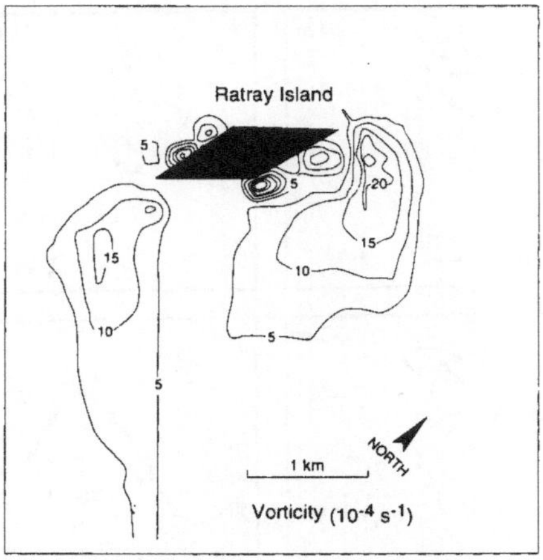

FIGURE 8.10. Predicted distribution of the vorticity just before high tide at Rattray Island. (Adapted from Falconer et al., 1986b.)

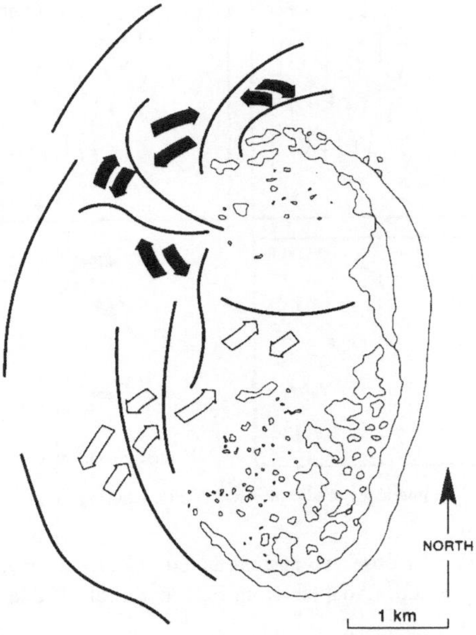

FIGURE 8.11. Topographically controlled fronts (lines) and their direction (arrow) of propagation with the tides at Bowden Reef. (From Kingsford et al., 1991. Reproduced with the kind permission of Springer-Verlag.)

five grid points representing an island (Black and Gay, 1987). This is simply because in the model diffusion occurs, both eddy diffusion in the equations of motion and numerical diffusion arising from numerical techniques. This diffusion

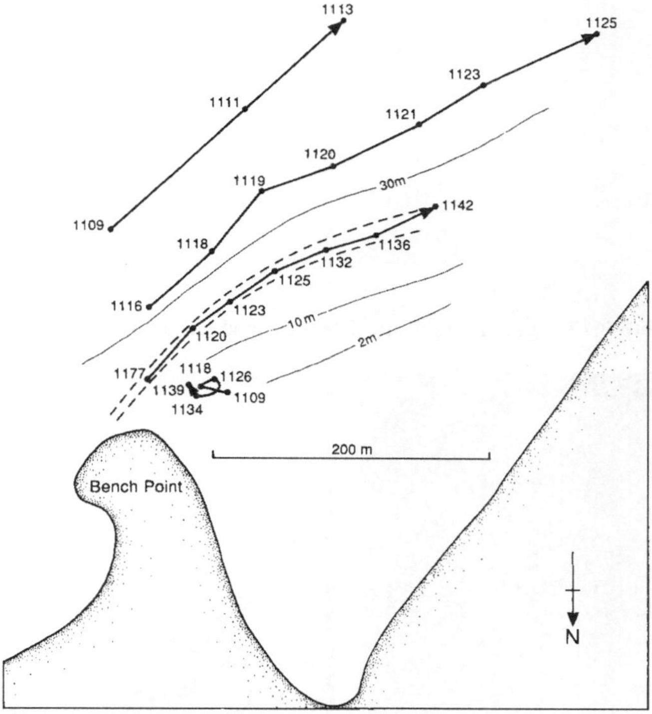

FIGURE 8.12. Bathymetry around Bench Point, trajectory of sail drogues, and (- - -) approximate location of the free-shear layer, at the flood tide. (From Wolanski, 1993. Reproduced with the kind permission of CSIRO Editorial Services.)

spreads the vorticity across vortex sheets, which are subgrid-scale phenomena not resolved by such models.

In shallow waters, friction permits quasi steady-state conditions to prevail in the lee of headlands and the vortex sheets do not fold onto themselves. An example of such steady shear layers occurs at Bench Point (Figure 8.12), a small rocky headland in the Whitsunday area (20.2°S). A free-shear layer is present for several hours during food tidal currents and is made apparent by two sharp lines on the water surface (Figure 8.12). The two lines were only about 5 m apart near the headland and separated slowly from each other with increasing distance. They were still readily visible 300 m behind the headland. Negligible temperature and salinity differences existed across this layer and the waters were vertically homogeneous in density. Curtain drogues (2 × 1 m at 2-m depth) were released on either side of the free-shear layer (Figure 8.12). The drogues in the free stream traveled rapidly with the tidal currents. The drogue downstream of the headland remained nearly stationary and was not entrained to the shear layer. This implies that the jet does not entrain across the shear layer, contrary to turbulent jet theory (Figure 8.13). The presence of eddies in the free-shear layer is also important. Transects

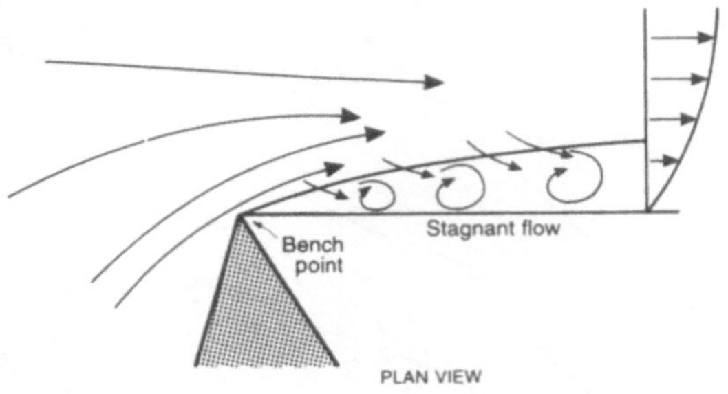

FIGURE 8.13. Sketch of the circulation at flood tide around Bench Point.

FIGURE 8.14. High-frequency acoustic soundings across the free-shear layer shed by Bench Point. (Adapted from Wolanski, 1993.)

of light transmissivity across the free-shear layer also show higher turbidity, due to a higher concentration of plankton and detritus in the eddy than on either side. Alldredge and Hamner (1980) found a similar aggregation of plankton in free-shear layers. High-frequency echo soundings across the free-shear layer showed a strong scattering in the shear layer, associated with a high concentration of zooplankton and detrital material (Figure 8.14), and the high concentration was present all the way to the bottom, a finding suggesting an organized upwelling mechanism in the eddies imbedded in the free-shear layer.

8.3 UPWELLING MECHANISMS AROUND CORAL REEFS

Details of the circulation around an island emerge from three-dimensional numerical models. Few such models have been applied to the Great Barrier Reef. The most detailed one is that of Deleersnijder et al. (1992), who solved numerically the fully nonlinear three-dimensional equations of motion using a staggered, finite-difference, explicit scheme which uses the finite volume approach (Nihoul et al., 1989). The hydrostatic assumption was made. Because the waters are shallow and well mixed, the momentum diffusion terms were simply parameterized using an eddy diffusion approach. The vertical eddy diffusion K_z was taken to be

$$K_z = 0.4\, u_* H\, (1 - 0.6\, z/H) \tag{8.15}$$

where u_* is the shear velocity, H the depth, and z the elevation above the bed. Horizontal diffusion is required to damp out small-scale computational noise and follows the procedure of Blumberg and Mellor (1987).

The model suggests that differences in horizontal velocity between the top and bottom layers (the bottom layer usually is located 1 m above the bottom) are typically <10% of the depth-averaged velocity. The distribution of the vertical velocity, expressed as the fraction $\Delta\sigma$ of the water depth that a particle of water travels vertically in 3 h, is shown in Figure 8.15a. In this figure, positive values indicate upwelling and negative values indicate downwelling. The model suggests that upwelling prevails in the eddy and downwelling along the island slopes. The predicted upwelling velocities appear to be too small, and it was suggested that this might be due to the model horizontal mesh size being too large (200 m). The location of the predicted downwelling zones explains the results of field studies by Hamner and Hauri (1981) showing accumulation of buoyant material and zooplankton upstream and on the side of islands (Figure 8.15b).

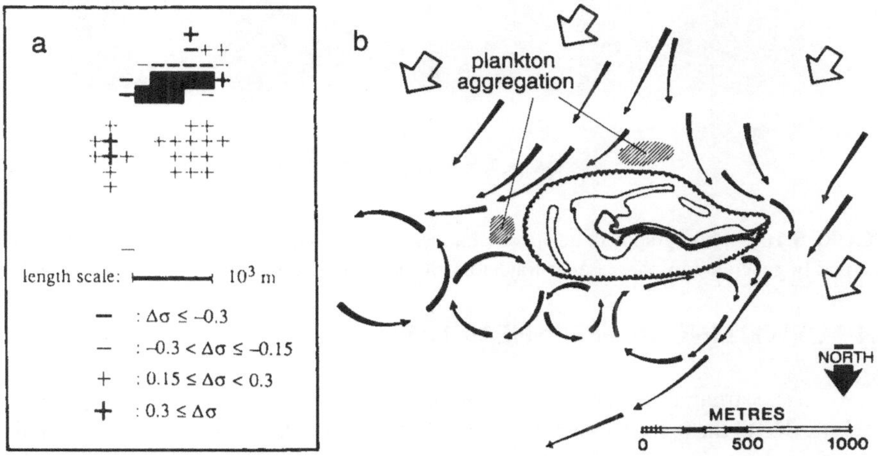

FIGURE 8.15. (a) Horizontal distribution around Rattray Island at flood tide of the upwelling parameter $\Delta\sigma$ (defined in the text; positive for upwelling, negative for downwelling) and (b) flow patterns around Pandora Reef (18.8°S) at (left) flood and (right) ebb tide. In b, the shaded areas show downwelling zones made apparent by aggregation of plankton. Note the encouraging similarity between observed and predicted downwelling zones. (Adapted from Deleersnijder et al., 1992; Hamner and Hauri, 1981.)

FIGURE 8.16. Photography of negatively buoyant particles entrained upward from the bottom in an eddy generated by the circular trajectory of the small submarine.

8.4 UPWELLING IN FREE-SHEAR LAYERS

The strong upwelling mechanism in free-shear layers (vortex lines) appears to be due to the presence of small-scale (5- to 10-m radius) eddies embedded in the free shear (Figure 8.13). The secondary circulation in the eddies was measured experimentally in a water tank, 1 × 1 m horizontal dimension, the depth varying between 30 cm and 1 m. Rotation was generated by a small (10-cm-long) model submarine attached to a surface buoy (Figure 8.16). The submarine moved in a circle of radius R (R ≃ 10 cm) at a period T = 1.3 s. This movement generated an eddy in the water with a period of about 10 s, and the circulation was observed well below the submarine. A strong upwelling was observed in the eddy, with the

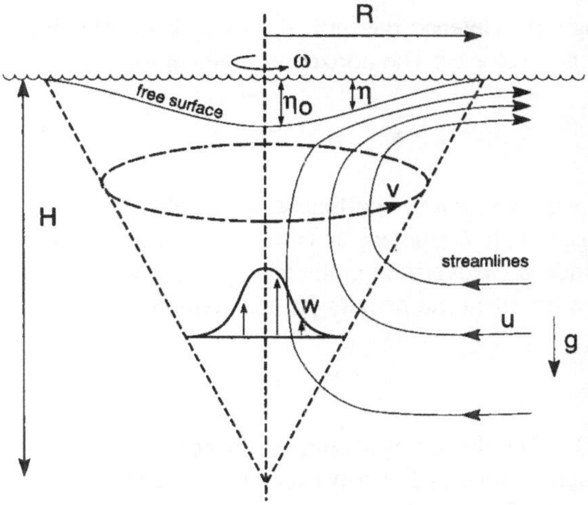

FIGURE 8.17. Sketch of the internal circulation in a small, intense eddy.

vertical velocity about equal to the horizontal velocity. The eddy radius R is found to decrease with distance from the surface (Figure 8.17). For $r < R$, where r is the distance from the center,

$$v = \omega r \tag{8.16}$$

where v is the radial velocity (Figure 8.17) and $\omega = 2\pi/T$ is the rotation rate. An upward velocity w prevailed, maximum at $r = 0$ and vanishing at $r = R$. This upwelling is intense, with $w_{max} \simeq v \simeq 5$ cm s^{-1}. The depth, H, of influence of the eddy is about 1 m. The upwelled water was deflected radially as it reached the surface.

These observations lead to the following scaling laws. Within the eddy, the centrifugal acceleration balances the pressure gradient.

$$v^2/r = g\, d\eta/dx \tag{8.17}$$

where η is the deflection of the free surface and g is the acceleration due to gravity. Assuming $\eta = 0$ at $r = R$, the maximum deflection of the surface occurs at $r = 0$, where

$$\eta_o = v^2/2g \tag{8.18}$$

The vertical momentum equation yields

$$w_{max} = \sqrt{(2g\,\eta_o)} \tag{8.19}$$

Upwelling is thus intense, the vertical velocity being of comparable magnitude to the horizontal velocity. The horizontal momentum equation yields

$$H \simeq v R^2 / \nu \tag{8.20}$$

where ν is the eddy viscosity coefficient. This yields $\nu = 5 \times 10^{-4}$ m^2 s^{-1}, a value 100 times larger than the molecular viscosity. This indicates that turbulent flow prevails, as indeed was verified in the laboratory experiments. Since the system functions as a turbulent momentum jet, the continuity equation yields

$$H = R/4\,\alpha \tag{8.21}$$

where α (≃ 0.025) is the turbulent entrainment coefficient. This yield H = 1 m. All these results agree with the laboratory experiments. The laboratory experiments also show that when h < H, where h is the water depth, strong radial flow toward the eddy center is generated with velocities u = v, in a thin benthic bottom boundary layer as sketched in Figure 8.5. In such situations, the bottom is rapidly denuded of material, as this material is carried upward and away in a thin vortex filament.

The internal circulation in the eddy is independent of the mechanism generating the rotation. Indeed, van Senden and Imberger (1990) found an upwelling in the vortex pair generated in a laboratory experiment by an unsteady jet. Onishi (1984) also found an upwelling in eddies embedded in a free-shear layer in a laboratory experiment.

Applying the scaling laws to the eddies at Bench Point where R = 5 m and v = 0.1 m s^{-1}, it follows that w_{max} = 0.1 m s^{-1}, H = 50 m, and ν = 0.2 m^2 s^{-1}. This suggests an intense upwelling within the eddy extending all the way to the bottom. At Bench Point, the lack of lateral entrainment in surface waters toward the free-

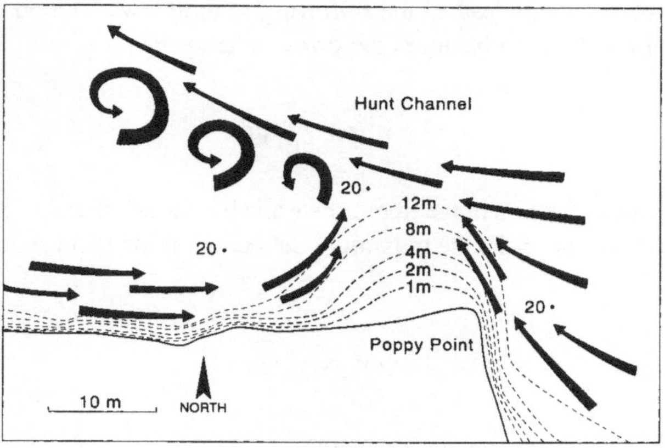

FIGURE 8.18. Current patterns around Poppy Point, with depth in metres. (From Alldredge and Hamner, 1980. Reproduced with the kind permission of Academic Press.)

shear layer (Figure 8.13) may be due to the radial velocity of the upwelled water on reaching the free surface (Figure 8.17). This implies that entrainment may still exist in deeper waters, near the bottom of the eddy.

In the lee of Bench Point, very shallow waters (<5 m deep) exist and friction prevents energetic flows downstream of the headland. At other headlands, the water is deeper and an eddy is formed in the lee of the headland. Field studies at Poppy Point, a nearby headland where the water depth exceeds 12 m (Figure 8.18), show that a recirculating flow exists in the lee of the headland. A free-shear layer is observed separating eddy water from the free-stream waters, and eddies are embedded in that free-shear layer. Zooplankton are found to be concentrated in the near-surface waters of the free-shear layer (Alldredge and Hamner, 1980). In turn, this aggregation brings in fish to feed preferentially in such zones, and sea birds also concentrate there.

8.5 SECONDARY CIRCULATION IN SMALL SHALLOW BAYS

Small bays along the coast and near coral reefs are common in the Great Barrier Reef. They are confined by two headlands, one on either side. Since one headland is always upstream of the bay, an eddy is present most of the time in the bay (Parnell, 1988). The eddy rotates alternatively clockwise and anticlockwise with the tidal currents (Figure 8.2a).

The fine sediment in these bays is not compacted but is biodisturbed and available for transport by secondary currents. Bioturbation in coral reef environments appears to be mainly due to the shrimp *Callianassa,* as well as other organisms such as gastropod snails (Roberts et al., 1981; De Vaugelas, 1985). Bioturbation can result in a complete turnover of the top 30 cm of sediment every 8 months. The loose sediment is ejected from burrow holes, forming small mounds readily erodible by currents in the range of 5 to 15 cm s^{-1}, and is redistributed by the weak currents in lagoons and embayments (e.g., Tudhope and Scoffin, 1984; Parnell, 1988).

The importance of the three-dimensional secondary circulation in moving sediment is apparent at Rattray Island, where fine sediment has been preferentially depleted from the sea floor by the eddy shed by the island at flood tide (Figure 8.19).

8.6 TIDAL JETS

Tidal jets are formed when strong currents are generated by the flow being restricted to a narrow reef passage (Figure 3.13). Two vortices are advected away from the passage with the developing jet. Free-shear layers are shed from each headland (Figure 4.16). This flow was studied in detail by Wolanski et al. (1988a) for the reef passage between Ribbon Reefs No. 3 and 4 (15.5°S; Figure 4.14). The reef passage is about 45 m deep and separates shelf waters about 40 m deep from

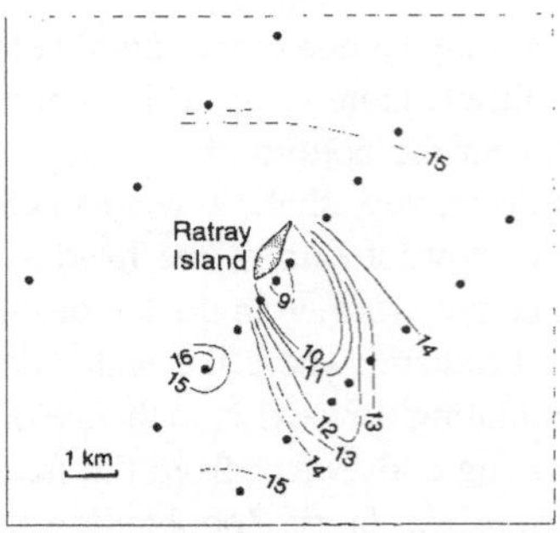

FIGURE 8.19. Percentage of bottom sediment of size <2.8 µm around Rattray Island. (Adapted from Wolanski, 1986c.)

the Coral Sea waters, where the water depth increases to 100 m at 1 km from the reef crest and over 2000-m depth further offshore. The water circulation was measured with radar-tracked drogues and with current metres deployed at 13 sites labeled A to M. The water circulation is forced by the reversing tidal currents through the reef passage. As the tidal currents turn to flood, the currents are too weak to separate from the reef and the flow field resembles potential flow (Figure 8.20a). When the currents exceed about 0.2 m s^{-1}, flow separation occurs and a tidal jet develops (Figure 8.20b). This jet grows rapidly in time and pushes at its leading edge a pair of counter-rotating eddies (Figure 8.20c). These eddies appear to have vorticity, i.e., the water velocity near the center increases with distance from the center. They are, however, embedded in a much larger vortex which has no vorticity, i.e., in the far field, the water velocity decreases with distance from the center. The situation is asymmetric, because of both earth rotation effects (Gidhagen, 1984) and the orientation of the passage at about 50° from the axis of the Ribbon Reefs.

The two-dimensional (depth-averaged) circulation was successfully modeled numerically by Wolanski et al. (1988a). The grid size was 166 m and the Manning coefficient was set equal to 0.025. The model suggested that the water that flows in the reef passage at ebb tide is drawn from outside the eddy generated at flood tide. Hence, the bulk of the water that enters the shelf at flood tide is not

FIGURE 8.20. Sketch of various stages of development of the tidal jet at Ribbon Reefs No. 3 and 4. (Adapted from Wolanski et al., 1988a.)

reexported at the following ebb tide, a finding supported by remote-sensing studies of tidal jets in Japan (Onishi, 1984, 1986). This phenomenon results from the pair of counter-rotating irrotational eddies having a self-induced velocity (Batchelor, 1967).

$$v_c = K/2\,\pi\,a \tag{8.22}$$

where K is the circulation and 2a the distance between the centers of the vortices. For the numerically predicted eddy pair, K can be estimated and thence $v_c = 0.1$ m s^{-1}, which compares favorably with that of 0.07 m s^{-1} obtained from the numerical model.

The model being depth-averaged neglects important three-dimensional effects. For instance, aerial infrared pictures of the eddies show the presence of a cold water patch, an observation suggesting an upwelling within the eddy (Wolanski et al., 1988a). The strength of this upwelling mechanism within an irrotational vortex moving over the sea floor is poorly known (Burgraff et al., 1971). van Senden and Imberger (1990) verified in laboratory experiments the existence of the upwelling mechanism in such eddies. They also proposed an analytical model whereby the vortex spin-up due to the roll-up of the free-shear layer is compensated by spin-down due to the Ekman suction in the eddies, so that ultimately the outflow structure becomes independent of initial conditions.

8.7 TIDAL CURRENTS IN A REEF MATRIX

A nonlinear, depth-averaged numerical model was used by King (personal communication) to study the tidal currents around Bowden Reef (19°S). The model domain, shown in Figure 6.5a, is vast, covering a 190-km stretch of the coastline and extending to the continental slope. The domain has three open boundaries, and these were specified by tidal heights predicted from tidal harmonic constants from tide gauge data. The mesh size is 2031 m. The predicted distribution at 2-h intervals of the tidal currents around Bowden Reef (Figure 8.21) is spatially heterogeneous. At 0400 h, the spatial heterogeneity is obvious in the lee of Bowden Reef and can thus be attributed to an island's wake effect behind Bowden Reef itself. At other times, such as 0200, 0800, and 1000 h, the heterogeneity is due to other reefs in the Great Barrier Reef matrix because the velocity field is spatially heterogeneous both upstream and downstream of Bowden Reef. At 1000 h, the undisturbed current upstream of Bowden Reef varied by a factor of 2 within 15 km along the east boundary. These predictions agree with the field data that show a large spatial gradient of the tidal currents in the far field of Borden Reef (Figure 4.5).

This finding brings into question the use of limited-area models to calculate the small-scale circulation near a coral reef. For instance, Wolanski et al. (1989) and Wolanski and King (1990) modeled the depth-averaged circulation around Bowden Reef in a 16.6 × 23.2 km domain. The mesh size was 386 m. The model was forced at the four open boundaries by observed tides and currents in the far field.

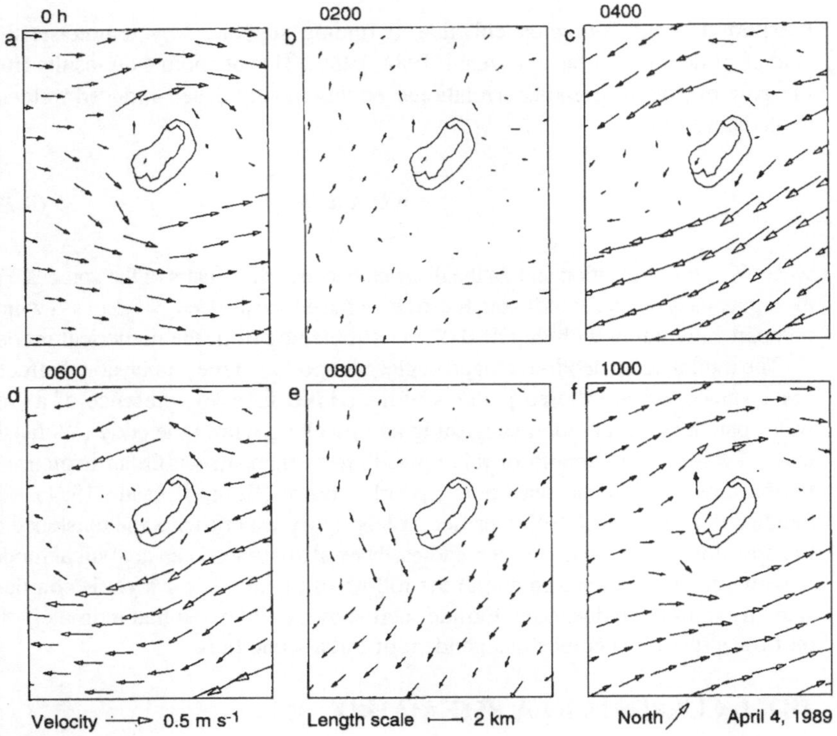

FIGURE 8.21. Predicted tidal currents around Bowden Reef on April 4, 1989, at 2-h intervals, using a large-scale model. The model domain is much larger and is shown in Figure 6.5.

The currents were assumed to be spatially constant along an open boundary because insufficient field data were available to specify the horizontal shear. An example of a prediction of the currents is shown in Figure 8.22. Eddies and shear layers are present but transient. The predictions are reliable near each of the four open boundaries, with a correlation between observed and predicted currents of about 85%. However, the correlation drops to 40% for sites in the center of the model domain where the reef is located (Oliver et al., 1992). This may be due to errors in specifying the currents at each of the four open boundaries as independent from each other and accumulating in the center of the domain. Gay et al. (1990) found similar results arising from errors in specifying the open boundary conditions in a model of the circulation around John Brewer Reef (18.6°S).

Thus, the water circulation around a coral reef cannot be reliably modeled by assuming that the reef is hydrodynamically isolated from surrounding reefs. This finding brings into question the reliability of all the small-scale models so far proposed of the circulation near coral reefs in the Great Barrier Reef matrix, as they all either neglect the sticky water phenomenon or assume a spatially uniform current forcing along the open boundaries of a limited-area model domain (Dight et al., 1990a, b; Black et al., 1990, 1991; Black and Moran, 1991; Scandol and James, 1992; Wolanski et al., 1989; Wolanski and King, 1990; Oliver et al., 1992).

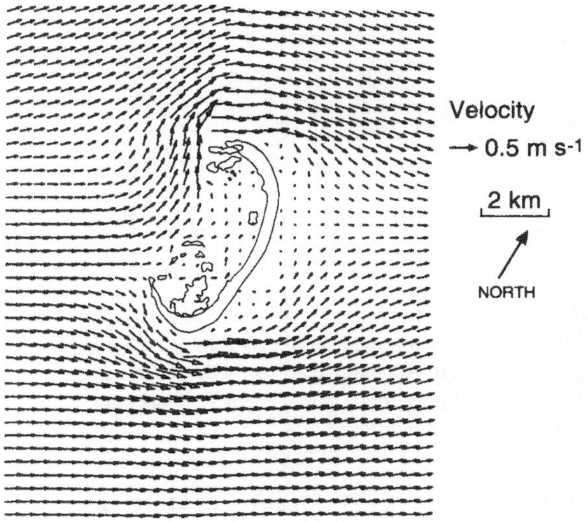

FIGURE 8.22. Example of the predicted current distribution around Bowden Reef using a limited-area model. (Adapted from Wolanski and King, 1990.)

This finding suggests that significant advances in reef circulation modeling require both a detailed data set of the currents through the Great Barrier Reef matrix, and the use of nested, nonlinear models forced simultaneously by the East Australian Current, the wind, and the tides. The use of nested models may enable one to progressively zoom in on the target reef. Several problems, however, remain unresolved, including how to parameterize the presence of free-shear layers, and the level in the cascade of models at which three-dimensional models should be nested with two-dimensional models.

The region of the reef where limited-area models show most promise is the reef flat. Figure 8.23 shows a time series of observed currents over the windward reef flat of Bowden Reef. Also plotted is the time series of the depth-averaged currents predicted by the two-dimensional, limited-area model of Wolanski and King (1990). High-frequency oscillations were also observed and are shown as error bars in the figure. Some high-frequency oscillations occurred in the absence of wave breaking. The model reproduced some of these oscillations. The predictions seem to be insensitive to the shear in the specified open-boundary currents. This finding may result from the simple dynamics prevailing over a reef flat, i.e., a balance between the sea-surface slope, wind stress, and bottom friction (Wilson, 1985). The sea-level surface slope can drive a current over the reef flat against the wind (Pickard, 1986; Hamner et al., 1988).

8.8 LARGE-SCALE CIRCULATION

In view of the unresolved interaction between tidal and low-frequency currents, and of the complexity of the reef-scale circulation, models of the large-scale

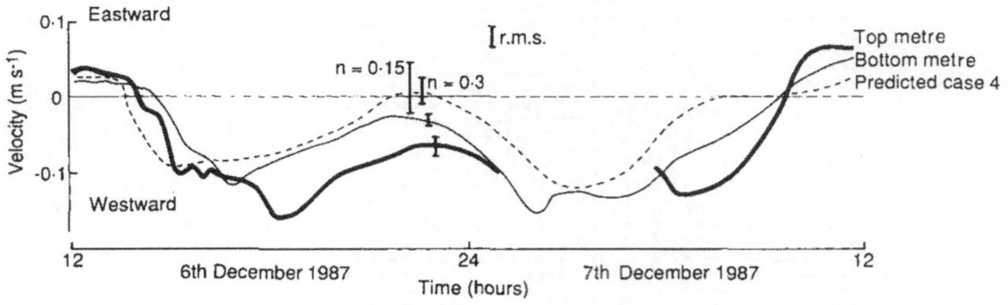

FIGURE 8.23. Time series plot of observed (thick line, top metre; thin line, bottom metre) and predicted (- - -) hourly averaged cross-reef currents over the reef flat at Bowden Reef. The error bars show the range of high-frequency fluctuations, both observed and predicted for various values of the Manning's n roughness coefficient. (From Wolanski and King, 1990. Reproduced with the kind permission of Academic Press.)

circulation in the Great Barrier Reef matrix have met with limited success. Bode et al. (1981), Bode and Stark (1983), and Bode and Sobey (1983) modeled the two-dimensional (depth-averaged) tidal and wind-driven circulation in the southern region. The mesh size was coarse, 5 nmil (≃ 9 km), so flows around coral reefs could not be included. Instead, reefs were modeled as weirs enhancing friction. The presence of the Great Barrier Reef had less influence on the solution than the specification of open-boundary conditions that were ill-defined in the absence of reliable tidal data offshore. Nevertheless, the model showed a large gradient of tidal amplitude and phase across the dense reef matrix. The results of the simulations showed the likelihood of unrealistic solutions emerging from inappropriate open-boundary conditions, and demonstrated the need for field data both to verify the models and to select the appropriate boundary conditions. This problem was also faced by Apelt and Richter (1985). Their grid size was 20 km cross-shelf by 40 km longshore. They modeled the reefs in the southern region as three 80-km-long barriers spread along the shelf break and separated one from the other by 80-km-wide gaps. Another barrier, 90 km long, was oriented longshore 60 km from the coast in the region of the Capricorn Group. A similar problem of ill-defined open-boundary conditions was faced by Andrews et al. (1983), who used a steady-state, linear model to study the mean circulation in the central region matrix and predict the movement of oil slicks.

Depth-averaged numerical models have been successful in reproducing tidal height and current dynamics in reef-free waters (the Capricorn Channel; Griffin et al., 1987). Some success was also obtained in reproducing the tidal dynamics in the central Great Barrier reef region in the area of least reef density (Andrews and Bode, 1988). The mesh size was 5 nmil (≃ 9 km). This model predicts tidal currents varying smoothly in the Great Barrier Reef, while both observations (Figure 4.5) and predictions from numerical models using a grid size four times smaller (Figure 8.21) show large spatial variability of the tidal currents.

A number of small-area models have been proposed, often by consulting engineers for local environmental impact studies, for the depth-averaged circulation in a number of embayments, such as Bowling Green Bay, Cleveland Bay off Townsville, and Trinity Bay off Cairns. These models are generally forced by the local wind and the tidal heights at the open-boundary conditions. With a few exceptions (e.g., King, 1992), most of these models have not been verified against long-term, time series current observations. None of them include wave-induced radiation stress (Signell et al., 1990). Most of them neglect the influence of the East Australian Current. With rare exceptions, most models neglect the influence of the tidal wetlands along the coast. Tidal wetlands profoundly influence the circulation in coastal waters in two ways (Wolanski et al., 1990; Wolanski and Ridd, 1990). First, a coastal boundary layer is formed comprising water moving back and forth with the tides between the mangroves and the shallow coastal currents. Little mixing occurs with waters further offshore because the coastal circulation is sluggish and the waters are shallow (Figure 2.11c). The water is thus trapped for long periods, up to 1 month, in such mangrove-fringed embayments. Second, at spring tides, the tidal prism is greatly increased by the mangroves, and coastal boundary water may be ejected as a tidal jet peeling off past headlands, such as has been observed at Cape Bowling Green (19.4°S) and Cape Cleveland (19.3°S; Figure 8.24). This mechanism may fertilize the offshore waters with nutrients outwelled from the mangroves.

A three-dimensional, nonlinear, baroclinic model was used by Wolanski et al. (1992c) to study low-frequency circulation in the reef-free Gulf of Papua. The model was forced along the coastline of Papua, New Guinea by river runoff, the observed low-frequency current in the Great North East Channel, the Coral Sea Coastal Current in offshore waters, and the local wind. Current-metre data were available for verification or for specification of open-boundary conditions, as well as CTD data. The model was successful in reproducing the stratification by brackish water throughout the Gulf and its long residence time (2 months) in the Gulf. It also had some success in reproducing the intrusion of brackish water from the Gulf of Papua in the Great North East Channel, provided realistic values of the currents through the Great Northeast Channel were used as a open-boundary forcing. The low-frequency circulation in the Gulf of Papua is sluggish. The current is oriented across-isobath on the inner shelf where river runoff and bottom friction dominate, along-isobath eastward or westward on the mid-shelf as a function of the wind, and along isobath southwestward under the influence of an eddy forced by the Coral Sea Coastal Current.

8.9 FLOW THROUGH THE SUBSTRATE

Movement of ground water through the reef substrate is believed to be very important from a management and scientific viewpoint. The problem of retention of ground water on coral cays has human implications for islands in Torres Strait, where ground water often constitutes the only local source of water in the dry season. Such is the case at Yorke Island, where the lack of water has forced a population overflow to migrate to nearby Sue Island. Ground-water flow may also have water quality implications in permitting the escape of effluents from septic

FIGURE 8.24. Aerial views of tidal jets peeling off from (a) Cape Bowling Green and (b) Cape Cleveland.

tanks from inhabited islands and coral cays to surrounding coral reefs. Groundwater flow may also measurably affect the nutrient budget of coral reefs (Andrews and Muller, 1983).

Notwithstanding all its presumed importance, flow through the coral reef substrate has received very little attention in the Great Barrier Reef. An experiment was attempted at the reef flat of Davies Reef to measure the speed of the ground water from a central borehole to a series of cased boreholes in a sampling grid around it (Oberdorfer and Buddemeier, 1986 and personal communication). Little of the tagged water reappeared at the wells, and horizontal velocities varied enormously, from 0 to 400 m d^{-1}. When drilling at Britomart Reef (18.3°S), S. Rhodes (personal communication) found that the drilling fluid was reappearing as a jet emerging from a crack in the substrate tens of metres away. These findings suggest that the flow is very rapid but largely limited to cracks and fissures, presenting to oceanographers the same challenges that face hydrologists studying the circulation of ground water in limestone. The coral substrate is so porous that Roberts et al. (1988) have been able to measure the wave-induced fluxes in pore waters at St. Croix, U.S. Virgin Islands. They showed that the ground-water fluxes of dissolved reactants enhance carbonate cementation and the biochemical transformation of shallow oxic pore-water environments into deeper anoxic environments. The pumping of large quantities of seawater through the porous system leads to precipitation of submarine cements (Marshall, 1985). These mechanisms probably explain why the reefs on the Great Barrier Reef are generally more consolidated near the windward margins than on the leeward margins (Hopley, 1990).

The porosity can exceed 40% in the uppermost Holocene veneer, which is 10 to 20 m thick in the Great Barrier Reef (Hopley, 1990). However, the porosity is spatially heterogeneous and patchy, and this patchiness has not been quantified.

Andrews and Muller (1983) studied the water flows past small coral outcrops in the lagoon of Davies Reef (18.8°S). They calculated the resulting pressure distribution around an idealized cylindrical outcrop, assuming a uniform porosity of the coral substrate. The flow within the substrate was calculated. Figure 8.25 shows predicted stream lines entering the substrate at the front and at the rear of the coral outcrop. An outflow occurs on the sides carrying nutrients out of the

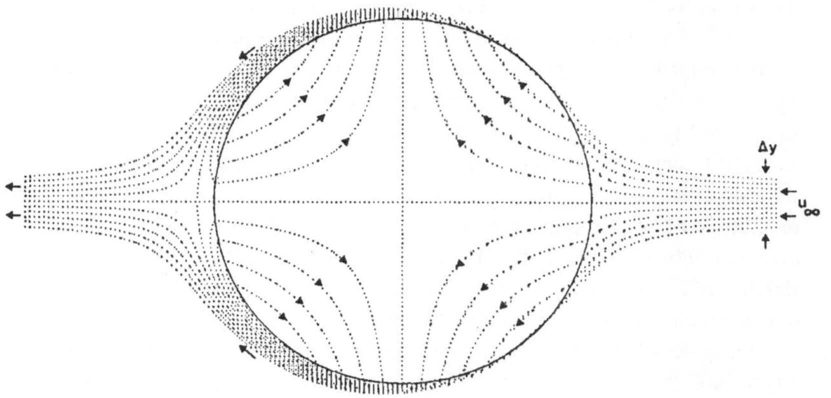

FIGURE 8.25. Predicted stream lines around and in the substrate of a small coral outcrop. (From Andrews and Muller, 1983. Reproduced with the kind permission of the American Society of Limnology and Oceanography.)

substrate. It was predicted that this circulation does not reverse direction when the prevailing far-field currents reverse direction by 180° with the tides. This is because the flow is pressure driven and the pressure gradient is proportional to the square of the speed of the tidal current.

8.10 BAROCLINIC CIRCULATION

The majority of the coral reefs of the Great Barrier Reef are located on the shelf where a barotropic circulation prevails, even if this circulation is three dimensional. On the shelf slope and in the Coral Sea, the waters are both deep and density stratified.

The circulation around coral reefs in deep waters of the Coral Sea is poorly known. The interaction of oceanic islands and reefs with strong currents may generate complex three-dimensional flows, isotherm doming, and boundary mixing, as has been suggested elsewhere in theoretical studies of an island in a steady flow by Hogg (1972, 1980) and Gordon and Hughes (1981) and in laboratory experiments by Baines and Davies (1980). Some field observations of isotherm doming in the lee of obstacles in a stratified ocean have been reported for Bermuda Island in the Atlantic (Hogg, 1972; Hogg et al., 1978), Aldabra and Cosmoledo atolls in the Indian Ocean (Heywood et al., 1990), in the Gulf of Lions in the Mediterranean Sea (Millot, 1979), and along the rugged South African east coast (Schumann et al., 1982). Remote-sensing images also suggest that vorticity shedding by headlands contributes to meandering of baroclinic coastal currents far downstream of the salient topography (Takahashi et al., 1981; Fu and Holt, 1982; Ikeda and Emery, 1984). These studies suggest that the currents past obstacles in a stratified ocean are unsteady and three dimensional. The dynamics of these flows are poorly understood. They may be biologically very important, as they enhance the productivity and biomass around the islands. This enhancement usually occurs in patches (Bakun, 1986; Heywood et al., 1990; Boehlert et al., 1992). Little is known of the dynamics of these patches.

A similar local enrichment was also observed in the Great Barrier Reef around Myrmidon Reef (18.3°S) on the shelf slope (Figure 4.9). The generation of eddies in the lee of Myrmidon Reef is most apparent in calm weather at neap tides when the East Australian Current is the dominant forcing. Current-metre data (Figure 8.26) and radar-tracked drogues (Figure 8.27) showed that the East Australian Current generated an eddy in the well-mixed layer in the lee of Myrmidon Reef. The circulation in the eddy was, however, unsteady. The flow in the bottom stratified layer did not rotate as did the eddy in the top layer. This eddy was accompanied by isotherm doming (Figure 4.11), which was superimposed on the internal tides.

Wolanski (1986d) proposed an analytical model of the circulation around Myrmidon Reef, assuming a stratified, two-layer ocean, and a prevailing steady current in the top layer. The vorticity shed in the top layer at the separation point generates an eddy in the lee of Myrmidon Reef. This eddy generates isotherm doming. The background rotation (the Coriolis term f) and the deflection of the interface contribute to the spin-up of the bottom layer because of the law of

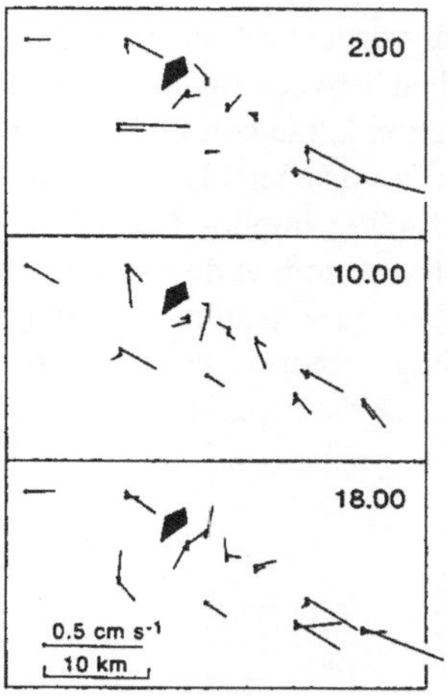

FIGURE 8.26. Typical velocity distribution around Myrmidon Reef at 8-h intervals. When two current metres are present on the same mooring, the velocity at the top metre is shown above that at the bottom metre. (Adapted from Wolanski, 1986d.)

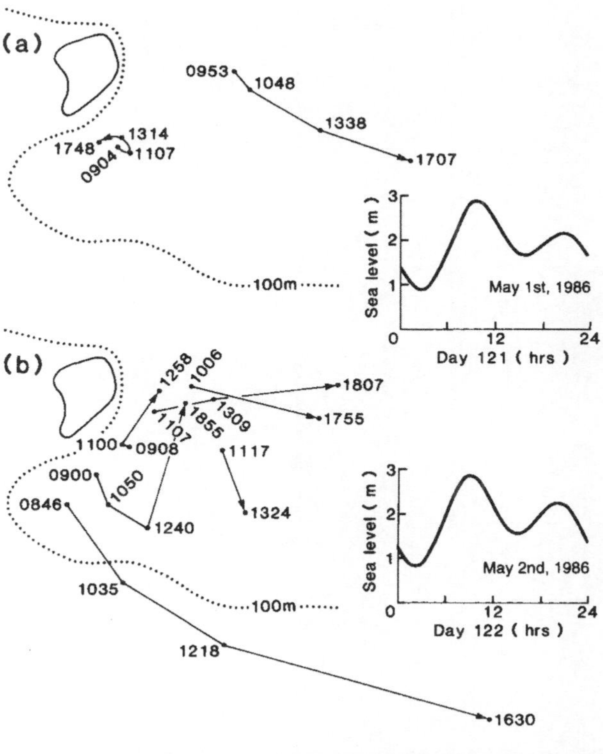

FIGURE 8.27. Trajectories of drogues with a parachute at 50-m depth on two successive days, around Myrmidon Reef. (Adapted from Wolanski, 1986d.)

conservation of potential vorticity. A steady state is assumed to prevail when a vorticity balance is reached between an input in the top layer from flow separation and a withdrawal by friction in the bottom layer. This yields an eddy size on the order of 25 km, which is considerably larger than the radius (≃ 1.8 km) of Myrmidon Reef. This finding implies that vorticity shed by Myrmidon Reef cannot be dissipated at the bottom at the same rate. Hence, vorticity is exported (advected) away from the area so that the current is disturbed far downstream of the reef. Vorticity shedding by reefs is thus likely to contribute to meandering, and possibly eddy formation, of the East Australian Current.

9 Mixing and Dispersion Around Coral Reefs

A number of water circulation processes generate patchiness, the dominant feature of coral reef oceanography. These can, broadly speaking, be divided into barotropic and baroclinic processes. These processes are described below.

9.1 BAROTROPIC PROCESSES GENERATING PATCHINESS

9.1.1 Lateral Trapping

The lateral trapping phenomenon (Figure 2.10) controls the shelf-scale dispersion processes in the Great Barrier Reef matrix. It operates at tidal frequency with the formation of eddies in the lee of coral reefs and islands. Lateral trapping behind islands and coral reefs is a dominant mixing mechanism enhancing the southward intrusion over 100 km of brackish water from the Gulf of Papua in the reef-studded Great North East Channel (Figure 3.10). This intrusion proceeds against a weak prevailing northward current of 0.05 m s^{-1} (Figure 5.13). With tidal currents $U_o = 0.6$ m s^{-1}, Equation 2.2 yields B = 91 m^2 s^{-1}, which is considerably larger than the horizontal diffusion values ($\simeq$ 5 m^2 s^{-1}) derived by Okubo (1974) for oceanic mixing at scales of 1 to 3 km. This value is also considerably larger than that (K = 2 m^2 s^{-1}) calculated from Equation 2.4 and due to vertical turbulent mixing and vertical shear dispersion. This result suggests that recirculating flows generated in the lee of coral reefs at tidal frequency provide a mechanism for enhancing mixing at intermediate scales of 500 m to 2 km in the cascade from large to small scales (Wolanski et al., 1984b).

9.1.2 Coastal Circulation

When horizontal mixing is weak, either because tidal currents are small or because of shallow water, streakiness can result. This is the case along the coast of the central region where a marked cross-shelf shear of the low-frequency current exists (Figure 6.5b) in the presence of the southeast trade wind blowing against the southward flowing East Australian Current. Tagged water particles move southward on the mid- to outer shelf but northward in coastal waters (King and Wolanski, 1990). As a result, the particles end up in a line centered around the zero velocity point. However, the line becomes segmented at spring tides by the strong tidal currents near the headlands (Figure 3.8).

9.1.3 Free Shear Layers

The presence of free-shear layers, which can also be called vortex sheets or barotropic fronts, has a profound influence on mixing near coral reefs. Free-shear layers are a common sight around reefs during periods of strong tidal currents, even in the absence of density stratification. The dynamics of these shear layers (see Chapter 8) are poorly known, but they are believed to be strongly three dimensional. Field data (e.g. Figure 8.13) show that these shear layers inhibit mixing between water masses on either side. Models of horizontal mixing in the presence of free-shear layers have been proposed by Signell (1989) and Signell and Geyer (1990). They used depth-averaged, two-dimensional models. Their grid resolution was adequate to explicitly calculate the presence of free-shear layers (Figure 8.7). They found that patches of contaminants in the vicinity of headlands are strongly deformed by straining associated with flow separation and the rolling of vortex sheets. The distribution of material becomes extremely patchy and streaky (Figure 9.1) for the water-borne particles entrained in the free-shear layers. The contaminants not entrained in the free-shear layers are essentially carried through with minimal horizontal mixing besides that due to the prevailing oceanic ambient turbulence at the small scales of the patch.

The key finding here is that near headlands and coral reefs, the patterns of transport and dispersion vary greatly over short-length scales and do not display the smearing tendency of a typical diffusive process. Thus diffusion is not Fickian and, in coarse scale models not explicitly calculating the presence of free shear layers, cannot be parameterized by the use a constant eddy diffusivity coefficient. This mechanism operates even in the absence of any buoyancy effects.

Similarly, the presence of free-shear layers in a tidal jet-vortex pair system (Figure 4.16) also generates patchiness and streakiness (Awaji et al., 1980; Awaji, 1982).

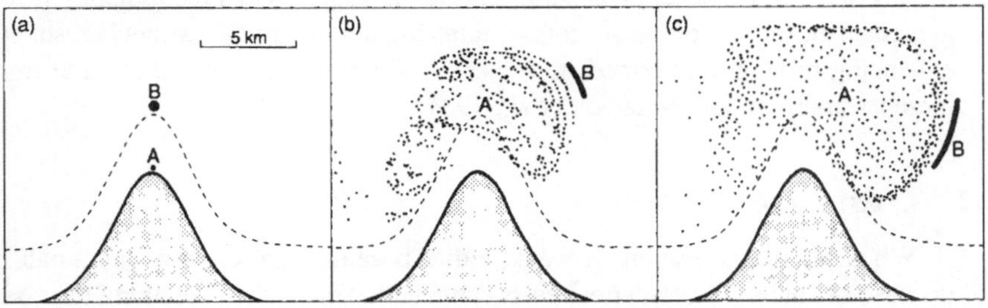

FIGURE 9.1. Numerically predicted distribution of a contaminant (a) released at points A and B near a headland in the presence of reversing tidal currents, after (b) six tidal cycles and (c) 12 tidal cycles. (Adapted from Signell and Geyer, 1990.)

Further, eddies embedded within a free-shear layer have a strong three-dimensional circulation, aggregating zooplankton near the surface (see Section 8.4). In turn, fish and seabirds aggregate in such areas.

9.1.4 Secondary Three-Dimensional Circulation

Eddies shed by the tidal currents around coral reefs and headlands have a secondary circulation with zones of downwelling where floating material, such as coral eggs, are aggregated in foam lines along the edges of an eddy (Figure 3.16; Wolanski and Hamner, 1988).

The downwelling near island slopes also results in the aggregation of plankton in downwelling zones (Figure 8.15). This mechanism does not require a completely recirculating flow, as these vertical motions are also present in a flow with a strong curvature near a solid boundary. This is the case of the barotropic flows upstream of reef passages. Zones of convergence and downwelling result (Figure 9.2a). An aggregation of plankton occurs at the convergence line (Wolanski and Hamner, 1988). Whales have been observed feeding on such aggregations.

Small waves breaking on an emerged reef also generate an undertow along the sloping surface of the reef. At some distance, the undertow is balanced by the surface wave- and wind-driven current toward the wall. Floating particulates, including coral eggs, are observed to be aggregated along a line that is parallel to, and about 1 m from, the coral reef.

Waves generate an upwelling mechanism in a spur-and-groove structure (Figure 9.2b), as was first observed by Munk and Sargent (1954) at Bikini Atoll. Roberts et al. (1975) studied this mechanism by releasing dye at a 33-m depth in a groove. They found that the water spread rapidly up the groove to reach the reef top in 1 or 2 min. This may be the largest upwelling velocity every reported worldwide. The mechanism of wave shoaling and breaking on a spur-and-groove structure has never been studied in detail. This mechanism has been little studied in the Great Barrier Reef, where the spur-and-groove structure is less well developed than in oceanic atolls with a dominant swell. The only study of this phenomenon on the Great Barrier Reef is that of Hamner et al. (1988), who used dye to follow the water movement near the windward sloping surface of Davies Reef (18.7°S). They found that the upwelling occurred only in the grooves, with a typical vertical velocity of 3 m min, so that groove water dyed at a 15-m depth crossed the reef flat 5 min later. Thus, the water crossing over the reef flat is not simply from the surface layer, but also comes from deeper water, from the depth at which the spur-and-groove system first develops. This mechanism upwells plankton from depth. In the daytime, most of this plankton is eaten by planktivorous fish forming a "wall of mouths", so that a considerable amount of detritus arrives on the windward reef flat in the daytime. At night, most planktivorous fish sleep, less plankton is eaten, and the concentration of upwelled plankton on the reef flat is higher than in the daytime (Hamner et al., 1988). Fish larvae also arrive in patches on reef crests (Victor, 1984), mostly at night, and this may be due to both

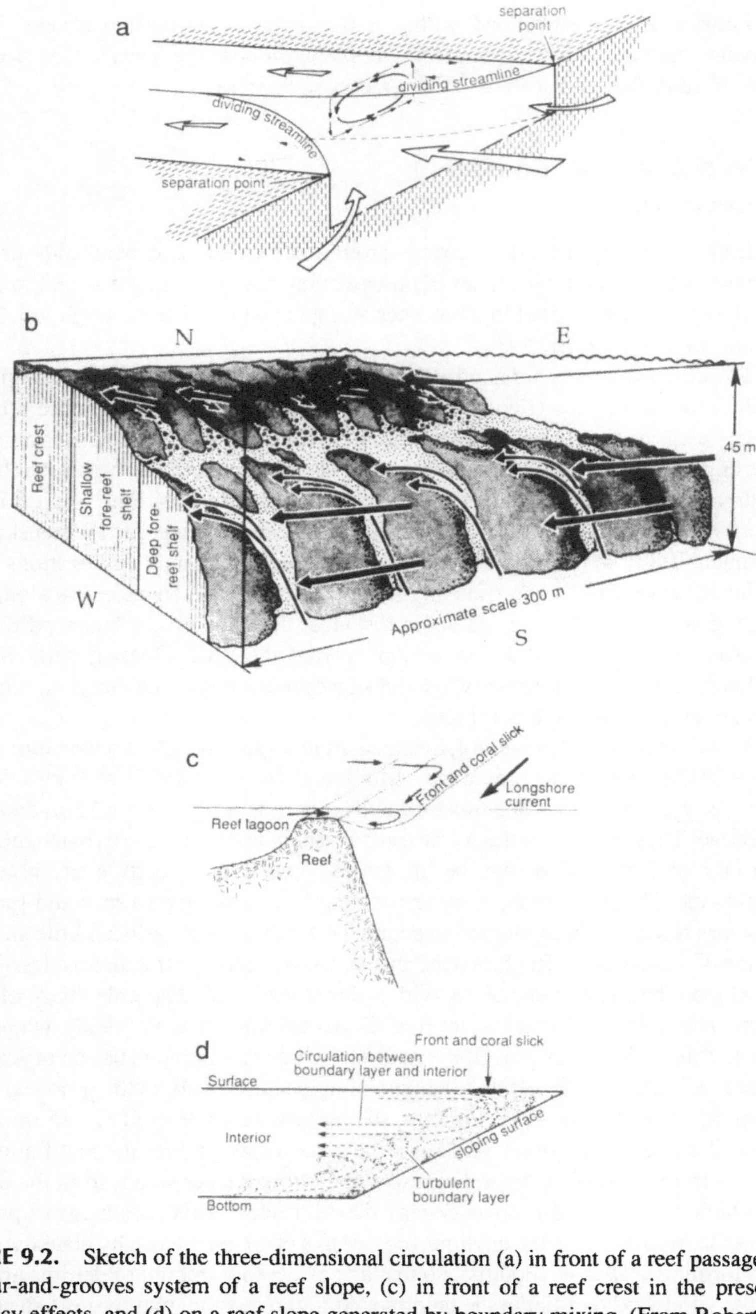

FIGURE 9.2. Sketch of the three-dimensional circulation (a) in front of a reef passage, (b) in the spur-and-grooves system of a reef slope, (c) in front of a reef crest in the presence of buoyancy effects, and (d) on a reef slope generated by boundary mixing. (From Roberts et al., 1975; Wolanski and Hamner, 1988. Reproduced with the kind permission of the American Association for the Advancement of Science.)

the "wall of mouths" in the spur-and-groove structure and to active swimming and vertical migration (Breitburg, 1989; Dufour, 1992).

9.1.5 "Sticky Waters"

The phenomenon of "sticky water" described in Chapter 8 can be expected to lead to the trapping of water-borne particles (eggs, larvae etc.) in areas of high reef density, while similar particles in surrounding areas of low reef density are readily advected away by the low-frequency currents. There is some evidence for this in satellite images of the chlorophyll distribution in the Great Barrier Reef (Gabric et al., 1990). Since the mechanism is tidally dependent, it will vary with the spring-neap tide cycle, trapping being more efficient at spring tides than at neap tides.

9.2 BAROCLINIC EFFECTS GENERATING PATCHINESS

Differences in buoyancy can be due to the positive buoyancy of water-borne particulates, e.g., coral eggs. They can also be due to temperature or salinity differences between two water masses.

This effect is readily illustrated in the mangrove swamps along the coast of the Great Barrier Reef, spring tides result in the inundation of the mangrove wetlands at flood tide. Mangrove litter, including leaves and seeds, float and are carried away at ebb tide. However, instead of being dispersed, they are commonly found in coastal waters aligned in long lines that can be up to a few hundred metres long and only a few mangrove leaves wide (Wolanski, 1992b). This outwelled detritus enhances primary and secondary productivity in coastal waters, and may in places reach the Great Barrier Reef (Alongi, 1990; Boto and Robertson, 1990; Robertson et al., 1988; Williams, 1988). These long lines of floating detritus are actually formed in the mangrove creeks, where they are readily visible. An aggregation is generated by the surface convergence (Simpson and James, 1986) toward the front in the thermal plume formed when swamp water returns to the creek at ebb tide, although the temperature difference is small (<1°C).

In the wet season, when salinity is not uniform, an aggregation can also be formed as a result of the surface convergence in the buoyancy-driven secondary circulation in mangrove-fringed estuaries (Figure 2.12).

A similar effect of surface convergence toward the front in thermal plumes is present in coral reefs and generates a "coral slick". The slightly warmer lagoon water flows over a reef flat, advecting coral eggs and forming a buoyant plume deflected parallel to the reef by the prevailing current on the shelf (Figure 9.2c).

Turbulent mixing is greatly enhanced by the roughness of the coral reef surface. Water layers flowing past the reef are mixed at the sloping boundaries (Figure 9.2d). The lateral intrusion of boundary mixed water thickens the thermocline, spreading the stratification both upward in the surface-mixed layer and downward in the bottom-mixed layer. Boundary mixing is particularly important in coral reef lagoons because the surface area of the solid boundary is greatly enhanced by numerous coral outcrops (Figure 1.6). The prevalence of boundary mixing is apparent from high-frequency echo soundings showing the presence of scatterers in coral reef lagoons. The scattering is probably due to aggregation of plankton in small temperature step structures. These scatterers are found in turbulent boundary layers about 2 m thick along the sloping surface of coral reefs (Wolanski, 1987).

Boundary mixing explains the large spatial differences in the nearly instantaneous vertical profiles of temperature around a small coral outcrop about 30 m in diameter in a lagoon (Figure 9.3). The prevailing current speed u at the time of sampling was about 0.05 m s^{-1}. The surface-mixed layer was about 14 m thick a few metres upstream from the obstacle. Mixing was intense along the boundary because a few metres downstream of the coral outcrop the surface-mixed layer was only 2 m thick. Boundary mixing results in an equivalent vertical eddy diffusivity.

$$K_z = P\,\delta\,K/A \tag{9.1}$$

where A is the plan area, P is the surface area where boundary mixing occurs, K is the eddy diffusivity in the boundary layer, and δ is the thickness of the turbulent boundary layer. The value of K can be calculated from boundary-mixing theory (Phillips et al., 1986).

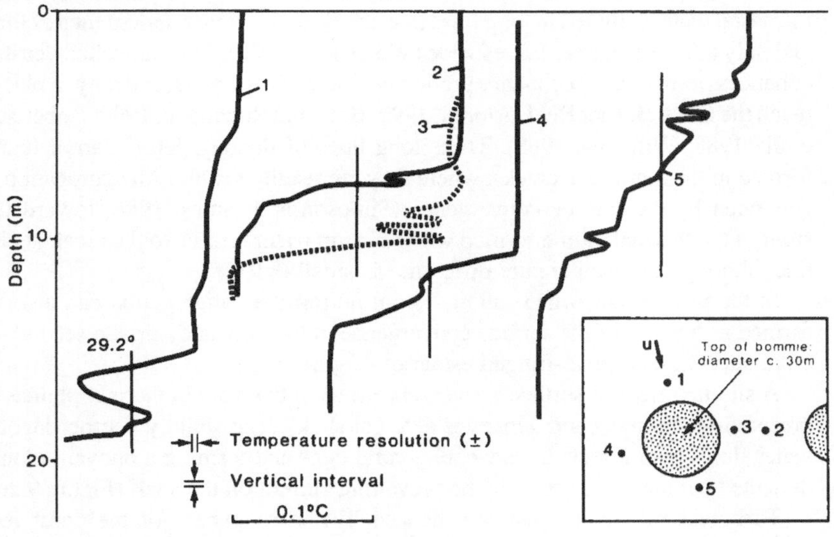

FIGURE 9.3. Profiles of temperature in the vicinity of a coral outcrop. (From Wolanski, 1987. Reproduced with the kind permission of the American Society of Limnology and Oceanography.)

$$\delta = (4\ K/N^2 \sin^2 \theta)^{1/2} \qquad (9.2)$$

where N is the stability frequency, ρ the density, g the acceleration due to gravity, z the vertical axis, and θ the slope of the solid boundary. The field data suggest that $\delta = 2$ m, $\theta = 60°$, and $N = 1.3 \times 10^{-2}\ s^{-1}$. Coral outcrops contribute to a large amount of mixing in coral reef lagoons. The lagoon of Davies Reef (18.7°S), for instance, has about 150 coral outcrops in its lagoon, so $P = 2.1 \times 10^4$ m. The mean lagoon depth H is 15 m. The time scale for vertical mixing of the lagoon is then $(0.7\ H)^2/K_z \simeq 2.7$ d. In practice, T will be even smaller because of mixing by wind. The small value of T implies that a strong temperature stratification is unlikely.

Boundary mixing generates an internal circulation in reef lagoons with a downwelling of surface water along the sloping boundary. A coral slick can form at the plunge point (Figure 9.2d).

Boundary mixing is an important mixing mechanism that enhances primary productivity in stratified shelf waters (Simpson et al., 1982). Boundary mixing may also be important in isolated coral reefs in stratified, Coral Sea oceanic waters, creating a subsurface chlorophyll maximum following the sloping pycnocline (Figure 9.4).

Boundary mixing is greatly enhanced by the rugosity of coral reef surfaces. It results in enhanced mass transfer of nutrients at the coral surface and in enhanced local productivity of the reef. The enhancement is several times that estimated

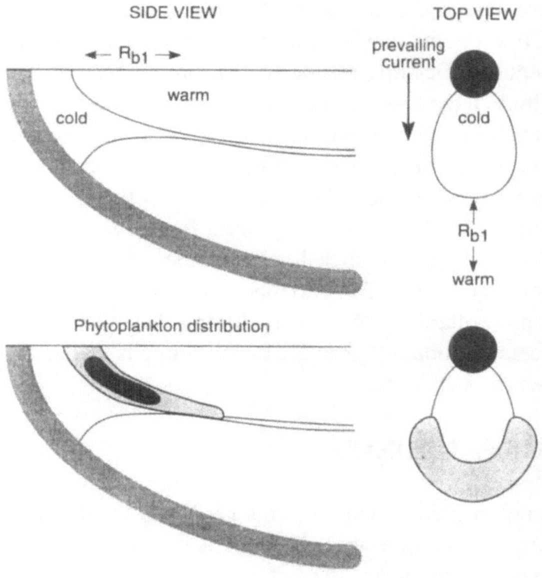

FIGURE 9.4. Currents and water masses distribution around an island in stratified waters, and phytoplankton distribution. (From Franks, 1992.)

from engineering studies of mass and heat transfer (Atkinson and Bilger, 1992; Patterson et al., 1991; Sebens and Johnson, 1991). One reason for this discrepancy may be the additional microscale turbulence generated by the motions of feeding tentacles, as happens through the moving arms of a filter-feeding crinoid (Colman et al., 1984).

9.3 OPEN WATER AGGREGATING MECHANISMS

A number of other aggregating processes have been studied in marine environments elsewhere, and presumably also apply to Great Barrier Reef waters but have not been studied.

Waves breaking on a beach generate rip currents, and these can aggregate diatoms (Talbot and Bate, 1987).

Towed video microscopy data from the continental shelf off Woods Hole, Massachusetts, revealed aggregations measuring less than 20 cm for copepods but not for nonmotile or asexual forms, a finding suggesting that these small patches are related to mating (Davis et al., 1992). In addition, aggregations on scales of 1 to 5 m were also found for most oceanic plankton, and these are probably hydrodynamically controlled.

Internal waves can also aggregate particles in surface slicks (Shanks, 1983). This mechanism probably occurs also at least in calm weather in coral reef lagoons where topographically controlled internal tides with a vertical excursion of 8 m in a 20-m depth have been observed (Wolanski, 1987). This mechanism probably operates efficiently on the continental slope, where internal waves of 100-m amplitude have been measured (see Section 4.3).

The downwelling mechanism in the corkscrew-like motions in wind-driven Langmuir cells is a well-known open-water aggregating mechanism (e.g., Barstow, 1983; Hamner and Schneider, 1986). Langmuir cells also exist near coral reef, where they aggregate *Trichodesmium* algae, fish larvae, presettlement reef fish, plankton, and coral eggs in slick lines (Hamner et al., 1988; Kingsford et al., 1991). Langmuir cells result from nonlinear dynamics involving waves and wind-driven motions (Pollard, 1976). By modifying the wave field but not the wind field, coral reefs presumably generate Langmuir cells that have different dynamics from those in open waters.

9.4 MODELS OF PATCHINESS

The presence of strong tidal currents generates shear zones and eddies, and lateral trapping results. The water-borne material, such as eggs and larvae, then leave the reef at tidal frequency in patches. Initially, lateral trapping thus enhances patchiness. Tidal trapping only enhances longitudinal mixing at time scales larger than the tidal period, provided that various patches are close enough that they can merge by mixing. In fact, such is generally not the case. The patches are usually too small and they are too far apart for each other to mix, since they are ejected from different areas of reef at different stages of the tidal cycle (Oliver et al., 1992). Thus, the primary effect of lateral trapping is to enhance patchiness.

In a spatially steady and homogeneous velocity field with a uniform horizontal diffusion coefficient (Fischer et al., 1979), the material concentration C within the plume can be calculated from the advection-diffusion equation, ignoring losses by settling and predation.

$$\frac{\partial C}{\partial t} + u\frac{\partial C}{\partial x} + v\frac{\partial C}{\partial y} = \frac{\partial}{\partial x}\left(K_x \frac{\partial C}{\partial x}\right) + \frac{\partial}{\partial y}\left(K_y \frac{\partial C}{\partial y}\right) \qquad (9.3)$$

where x and y are the horizontal axes with the origin being at the source (the reef), u and v are the horizontal velocities, t is the time, and K_x and K_y are the horizontal diffusion coefficients. For a spike release, e.g., following mass spawning, a number N_o of larvae are released at the reef and the concentration of larvae at any point (x, y) downstream of the reef and at a later time t is given by

$$C = \frac{N_o}{4\pi}\left(\sqrt{K_x K_y t}\; e\right)^{-1-(x-ut)^2/4K_x t} e^{-(y-vt)^2/4K_y t} \qquad (9.4)$$

The larval concentration would then form a smooth plume, as proposed by Williams et al. (1984) and as sketched in Figure 9.5a for the case of Bowden Reef (19°S). In this figure, the plume boundary was determined from field studies of the spread of radio-tracked drogues (Wolanski et al., 1989). Field studies revealed

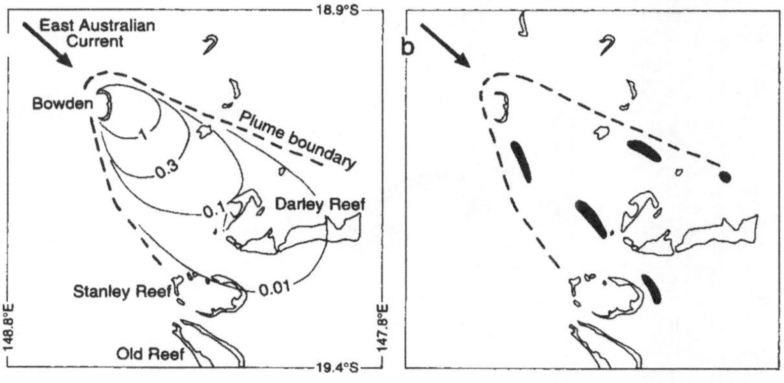

FIGURE 9.5. Sketch of the distribution of coral larvae density after mass spawning and flushing from Bowden Reef in November 1986, from (a) steady-state oceanic dispersion models assuming Fickian diffusion and (b) observations.

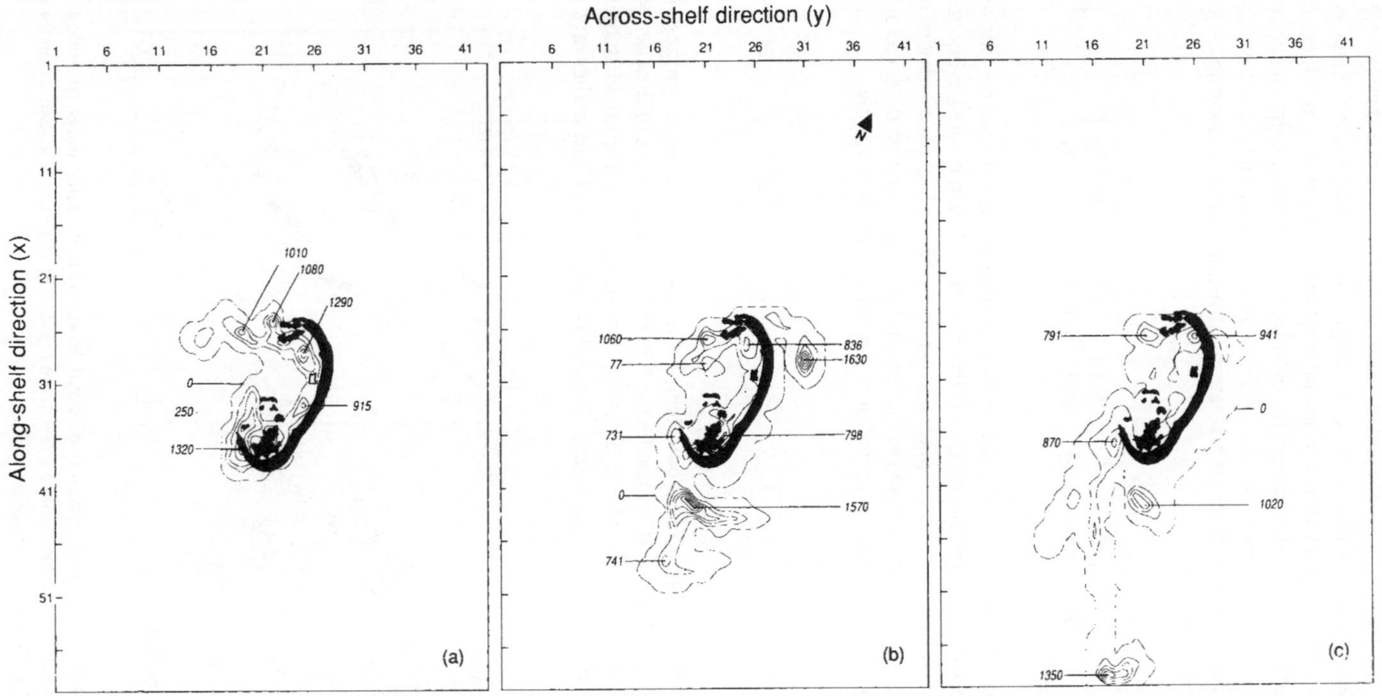

FIGURE 9.6. Predicted distribution of larval concentration (in number of larvae per cell) around Bowden Reef, (a) 1 d, (b) 2 d, and (c) 3 d after uniform release of larvae over Bowden Reef on December 10, 1987. (From Oliver et al., 1992. Reproduced with the kind permission of Pergamon Press.)

considerable patchiness within the plume, as sketched in Figure 9.5b, with zones a few hundreds of metres apart showing a difference factor of 10 in larval concentration (Oliver et al., 1992; Willis and Oliver, 1990). To model this situation, Oliver et al. (1992) used a depth-averaged, two-dimensional, advection-dispersion model to predict the larval distribution following mass coral spawning at Bowden Reef. The hydrodynamical model was calibrated against an extensive current-metre data set (Section 8.7). They predicted considerable patchiness in the larval distribution (Figure 9.6) within the general path of the larval plume that was drifting away with the East Australian Current. In this simulation, the patches were mainly formed by the ejection at tidal period of lagoon waters from, alternatively, the northern and southern regions of the reef. The model mesh size (386 m) was sufficiently small that the model was able to predict eddy shedding, but was too coarse to explicitly include free-shear layers. The patches predicted by the model were due to the lateral trapping phenomenon, not the rolling of free-shear layers. A comparison of the observed and predicted distributions of larvae showed that the model considerably underestimated the patchiness.

To show the likely importance of the parameterization of free-shear layers in dispersion processes in the Great Barrier Reef, Wolanski (1993) modeled the circulation around a small (1000 m long × 200 m wide) rectangular island in 18 m of water. The model grid size was 200 m. A steady far-field current was assumed (Figure 9.7). A passive, water-borne contaminant was introduced up-stream of the island, and the contaminant plume was calculated by the advection-

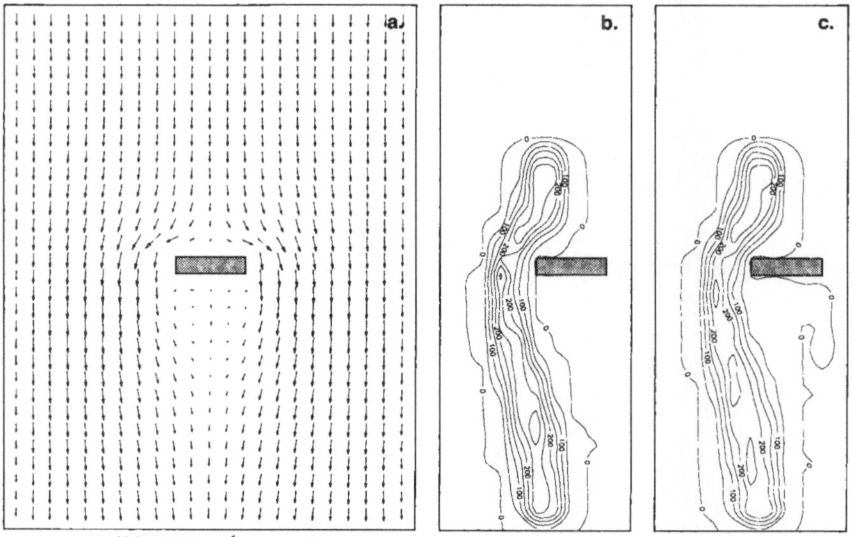

FIGURE 9.7. (a) Predicted steady-state circulation around a reef, and predicted larval concentration (in number of larvae per cell) in the (b) presence and (c) absence of a free-shear layer. (Adapted from Wolanski, 1993.)

diffusion model. The simulation was done for two cases, i.e., in the presence and in the absence of a subgrid scale, free-shear layer emanating from the separation points at the tips of the island. The presence of this subgrid scale layer was parameterized by assuming zero horizontal mixing across the free-shear layer. Elsewhere, a uniform horizontal eddy diffusion coefficient of 3 $m^2 s^{-1}$ was assumed. In the absence of a free-shear layer, the horizontal eddy diffusion coefficient was 3 $m^2 s^{-1}$ everywhere. The predicted contaminant plume is shown in Figure 9.7b and c. Comparing the two plumes, there are no differences upstream, but downstream there are major differences. No contaminant is trapped behind the island in the presence of a free-shear layer, while some trapping occurs in the absence of a free-shear layer. The two results are thus fundamentally different in terms of trapping behind the island, implying that it is vital to parameterize the presence of free-shear layers in coarse grid models of advection and dispersion processes in the Great Barrier Reef.

10 Managing the Great Barrier Reef

10.1 THE IMPACT OF MAN

The impact of man is only a very recent phenomenon, and the time scales of this impact are insignificant when compared to those of the reef's history. About 20 million years ago, the northernmost part of the Australian continental shelf moved into waters warm enough that coral growth was possible (Hopley, 1982, 1990). The tectonic plate has continued moving north at a rate of 4 to 8 cm per year. However, much of the Great Barrier Reef is probably not more than half a million years old. There were during that time large sea-level fluctuations, and much of the time the present Great Barrier Reef was not in the water but exposed. During the shorter periods of immersion, when the sea level rose, corals recolonized the area.

The Great Barrier Reef was an unspoiled area of great natural wonder and beauty when it was found by Europeans a little more than 200 years ago. Their impact in that short time has been marked and has occurred in a vacuum of scientific understanding of the physical and biological oceanography of the region. If the integrity of the ecosystem is to be protected for its conservation values, tourism, and economic resources, we now require a much greater scientific understanding of the Great Barrier Reef.

Man's impact started at the local scale, probably by the release of goats on islands to provide food for sailing ships. The goats devastated the native vegetation. Along the coastal plain, forests were felled to clear the land for agriculture and settlement. At times, this clearing was so dramatic that Ratcliffe (1947), for instance, described fires burning all year round as rain forests were cleared over the Atherton Tablelands near Cairns (16.9°S). The remaining forests were not fully protected until the late 1980s. Near coastal communities, navigation channels were dredged, and the dredge spoil dumped at sea, sometimes near coral reefs. Some stands of mangroves were cleared. Fishing was largely unrestricted until the late 1970s. Farming and pastoralism developed — thus accelerating soil erosion. Insecticides were sprayed on farm and plantation crops, and fertilizers added to the soil. Roads were constructed and the soil dumped in nearby creeks and over wetlands running in coastal waters. Sand was dredged from some beaches. Overburden and mine tailings were discharged in rivers, such as the tin mining along the Herbert River. While this practice has stopped in Australia, it is still common in Papua New Guinea and in Irian Jaya, with a potential pollution risk for reefs in Torres Strait.

On the coral reefs in coastal waters, environmental degradation has occurred, since hard corals have a low tolerance for fine sediments. Hopley (1990) reported that the construction of the Cape Tribulation Road in coastal rain forest in the 1980s has increased 22-fold the sediment inflow to adjacent coastal waters of the Great Barrier Reef. In another example, pictures (Figure 10.1) of the western beach of Magnetic Island (19.3°S), located near coastal developments and a dredge spoil site, show that the mangroves have colonized an area that was a sandy beach 50 years earlier. This beach was then fringed by a live coral reef, which is now largely a mud bank. Other old photographs show ocean-going vessels in coastal rivers that are now only knee deep. The late Sir Maurice Yonge (personal communication) described how the hard corals he found at Low Isles (16.4°S) during the British Museum scientific expedition in 1927 and 1928 were largely replaced by fine sediment and soft corals when he revisited the site 60 years later. The potentially lethal box jellyfish appears to be much more common in coastal waters now than 50 years ago, according to eyewitnesses, and a number of reasons may be possible, including the commercial killing of turtles, a predator of the jellyfish. There are numerous such examples, documented in qualitative and quantitative forms, of the environmental perturbation along the coast. Until recently, little has been attempted to quantify the overall impact of anthropogenic changes on the tropical coastline, and such an exercise is further complicated by the lack of understanding of the extent and scale of natural variability (biological and physical) in most coastal systems.

The impact of man has also been felt offshore, but is often less evident because of the remoteness of the region. For example, at Green Island (16.8°S), nutrient enrichment by sewage has led to a growth of seagrass beds from about 1900 m^2 in 1945 to 130,000 m^2 in 1978 (Hopley, 1990). The seagrass has mobilized sand, and this has enhanced erosion of the island. At Heron Island (23.4°S), a navigation channel was dredged. This has captured sand and increased erosion of the beach. In addition, dredge spoil from the navigation channel was dumped on the beach in September to November 1987 (Figure 10.2). This fine sediment was quickly eroded by waves. The resulting silt plume has polluted the marine environment and killed some coral communities. The construction of marinas and an airport on islands has led to some local pollution problems. At Hamilton Island (20.2°S), for instance, I found that the extension of the airport runway into the coastal waters interacted with strong tidal currents and generated at ebb tide a zone of turbid, recirculating flow in which pollutants from the harbor (plastic bags and bottles) remained for several hours (Figure 10.3), notwithstanding 4-knot currents a few metres offshore and a 3-m tide.

Piorewicz and Saunders (1990) reported that about 100,000 m^3 of sand was dredged in 1970 to 1972 from the beach at Yeppoon (23.2°S) on the Capricorn

FIGURE 10.1. View of Cockle Bay, Magnetic Island off Townsville, in (top) 1937 (low tide, later afternoon) and in (bottom) 1991 (mid-tide).

Coast, under the belief that the beach would replenish itself. Twenty years later, the beach still has insufficient sand, erosion is severe, and engineering works are needed to protect the coast (Figure 10.4). Recently aquaculture has been developing along the coast, resulting in the removal of some mangroves while constructing ponds and access roads or laying pipelines (Figure 10.5), and in some cases in eutrophication of coastal and intertidal waters by waste discharge.

FIGURE 10.2. Aerial view of Heron Island, its dredge channel, and the sediment spoil on the beach.

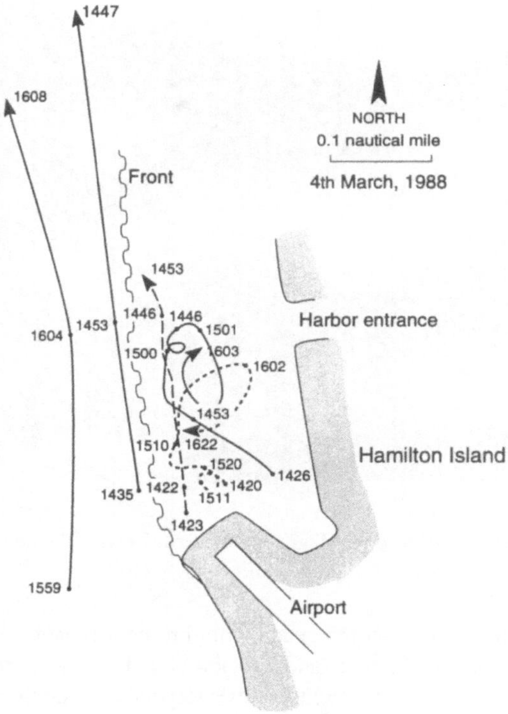

FIGURE 10.3. Trajectories of drogues around the airport runway extending in coastal waters at Hamilton Island, in the Whitsunday Island area, March 4, 1988. Local time is used.

FIGURE 10.4. Coastal protection structures at Yeppoon have become necessary following sand dredging offshore.

FIGURE 10.5. Clearing of mangroves for water pipes for aquaculture ponds located above the tidal limit.

Economic downturns have halted midway some construction projects on the coast, scarring the coastline and polluting coastal waters with fine sediments (Figure 10.6).

Gibbs (1993) reported heavy metal pollution in Townsville Harbor (19.3°S). Olafson (1978) found small but measurable levels of organochlorine pesticides in

FIGURE 10.6. Photograph in 1993 of an uncompleted marina at Nelly Bay, Magnetic Island off Townsville (19.2°S), 3 years following bankruptcy.

hard corals, fish and mollusks over the outer reefs off major rivers. There has been no comparative studies since, although agricultural activities have continued to increase. Bell (1990) reported that discharges from agricultural activities have continued to increase and that discharges from agricultural, industrial, urban, and tourism developments are now significantly affecting the water quality of the Great Barrier Reef. Using data from the British Museum scientific expedition to the Great Barrier Reef in 1927 to 1928, he also argued that levels of phosphate and phytoplankton have increased significantly over the past 60 years.

Craik (1990) reported that coastal authorities are not equipped to deal with anything more than a small, 1000-ton oil spill, and then only in the immediate vicinity of the main cities.

The growth of harbors, which have to cope with increasing shipping volumes in the region, presents problems with both the turbidity generated by dredgers and the disposal of dredge spoil at sea, sometimes only a few kilometre from coral reefs. There has been no attempt yet to dredge behind curtains or to release the muddy overflow on the bottom, so as to minimize spreading the fine sediment. Environmental studies are still largely restricted to short-term impact during and immediately following dredging and spoil dumping. In some cases the proximate criterion for the environmental impact study of dredge spoil is not the level of stress to the coral but simply whether or not the coral dies during dredging and spoil dumping.

While economic developments such as those given above present local management problems, outbreaks of the crown-of-thorn starfish *Acanthaster planci* over the last decade have generated major biological disturbances affecting in various intensities large parts of the Great Barrier Reef. During the last 12 years, the main effects of the starfish have been within the central Great Barrier Reef between 14 and 20°S. Statistical analyses of the shallow water (depth < 20 m) survey data suggest that outbreaks appeared to begin on reefs in the proximity of 16°S and then moved to the south and north of that region. By 1991, the more dominant southerly movement had covered a distance of almost 500 km. The outbreaks have caused substantial damage to corals. Surveys conducted over the last 7 years have indicated that approximately 57% of reefs with outbreaks had moderate to high coral mortality (>50%) over at least one third of their perimeters. In some reefs the devastation by the starfish has to be seen to be believed. For instance, in some parts of Charity Reef (19.3°S), the corals have been so thoroughly eaten out by the starfish, and the remaining substrate so much attacked by borers, that the coral skeleton literally appears to crumble and turn to rubble. Massive *Porites* corals have even been destroyed by crown-of-thorn the last 12 years (Figure 10.7), something which may not have happened in the past, judging by the uninterrupted continuity of growth as recorded in coral rings. There is evidence that the fast-growing colonies of Acropora and Montipora may recolonize a reef within 12 to 15 years, assuming no additional disturbances or outbreaks in the meantime. However, it may take populations of the slower growing, massive *Porites* coral in excess of 50 years to recover.

It is not known why outbreaks of crown-of-thorn starfish have occurred in such intensity in recent times. Surveys were unfortunately largely limited to the top 20 m of the water column. Several hypotheses have been put forward to account for primary (i.e., initiating) outbreaks; these include the adult aggregation hypothesis, the larval recruitment hypothesis, the terrestrial runoff hypothesis, and the

FIGURE 10.7. Massive *Porites* coral attacked by crown-of-thorns starfish.

predator removal hypothesis. For instance, in the predator removal hypothesis, little is known of the impact of prawn trawling that routinely discards as trash the majority of the catch, which is composed of small and large fish (Figure 10.8) that may be important predators of the crown-of-thorn starfish larvae. It may be that the cause of primary outbreaks may be far more complex and involve a combination of factors, possibly both natural and man-induced.

Most of the devastation to corals appears to have been due to secondary outbreaks, i.e., starfish outbreaks in reefs downstream and some time after the primary outbreaks. It has been suggested that these outbreaks are due to the advection and dispersion in interreefal waters of starfish larvae from reefs with primary outbreaks. However, because surveys were largely limited to depths < 20 m, little information was gathered on whether secondary outbreaks may also be due in part to the reef-to-reef migration of adult starfish leaving a devastated reef to avoid starvation. There is at least one visual observation, by a nonscientist, of an aggregation of adult starfish on the sea floor in interreefal water at a 40-m depth. This may have important management implications because if the adult starfish did migrate, they could be destroyed and the infestation managed, as sea urchins were controlled in California coastal waters following their infestation of kelp beds. No data or models are available to predict whether such measures would be effective for controlling crown-of-thorn starfish.

Lack of understanding of the causes of outbreaks has restricted effective management of the problem on the Great Barrier Reef. Up to the present time, control of outbreaks has been tactical in nature, being undertaken in very small areas which are of importance to tourism or science.

FIGURE 10.8. "Trash" fish discarded by prawn trawlers.

10.2 PHYSICAL OCEANOGRAPHY AS A TOOL FOR MANAGEMENT

Management of the huge and topographically complex Great Barrier Reef should be based on a sound understanding of key aspects of the water circulation and the variability of the physical and biological components. Water circulation controls the advection and dispersion of eggs and larvae from fish, corals, and crown-of-thorn starfish when they are in the pelagic dispersal phases. Water circulation also controls the transport of pollutants such as nutrients from sewage and agricultural runoff, pesticides, and mud from dredging. The dynamics of the water column also exerts forces on man-made structures and controls important processes such as erosion and siltation. This in turn affects turbidity and the marine ecology.

Because of the complexity of the topography, there is no single description applicable for the whole Great Barrier Reef. To help management, processes need to be understood and this knowledge applied on a case-to-case basis. This approach requires oceanographic field data and a combination of fluid mechanics, hydraulic, and engineering studies. A reliance on linear mathematical models without field verification may be misguided.

A sound understanding of the transport of water and particulates in the matrix of the Great Barrier Reef still eludes us. It is unrealistic at present (Wolanski, 1993) to expect to model reliably over several weeks the fate of water-borne, neutrally buoyant particles, such as *Acanthaster planci* larvae, because of unresolved problems in the specification of open boundary conditions and in parameterizing free-shear layers. The discrepancies between observed and predicted currents accumulate and become dominant after a few weeks of advection and dispersion. The more complex case of buoyant particles, e.g., coral eggs and oil, is even less amenable to prediction, given the present lack of knowledge of how to parameterize small-scale, topographically-induced, three-dimensional currents and the resulting aggregation mechanisms.

To improve our understanding and answer pressing management questions, a number of initiatives are needed. Field data need to be gathered on the complex flows through the Great Barrier Reef matrix. Studies in areas of high reef density show significant blocking of the current, possibly modulated at the spring-neap tidal period. Future research should concentrate on key fluid mechanics phenomena governing mixing processes around coral reefs. Complex numerical problems must also be overcome in the nesting of two- and three-dimensional models in order to calculate the interreefal circulation. The models to be developed for management must be nonlinear and the forcing must include, simultaneously, at least the three dominant forcings: the East Australian Current, the wind, and the tides, including the neap-to-spring tidal cycle. Also, the grid must be small enough to reproduce the interreefal circulation, even in the coarse scale models.

In addition, field experiments need to be carried out to be able to understand, and hence to parameterize in numerical models, the small-scale, topographically induced circulation leading to non-Fickian diffusion, aggregation, patchiness, and chaos. These studies should concentrate on both the barotropic and the baroclinic processes.

The fate of patches of larvae drifting through the Great Barrier Reef matrix has never been addressed and needs attention. Detailed multidisciplinary, process-oriented field studies, with a sound oceanographic component, are needed in order to provide data for model calibration.

Present knowledge of hydrodynamics makes difficult reliable predictions of oil spill trajectories in the Great Barrier Reef. Research on likely oil dispersion processes in a complex topography is clearly warranted.

Field studies (e.g., Oliver et al., 1992) suggest that coral recovery is initiated through the influx of new larval recruits which are advected from reefs further upstream. Studies of this phenomenon were initiated in the late 1980s, but have now been halted for budgetary reasons and clearly need to be initiated again.

The dearth of studies of suspended sediment dynamics in the Great Barrier Reef precludes a clear assessment of the impact of human activities on the Great Barrier Reef. Human activities increase sediment and pollutant input to the Great Barrier Reef. The sediment cohesive properties probably ensure that the bulk of the pollutants is attached to the sediment. It is thus likely, but has never been investigated, that the organochlorine pesticides found by Olafson (1978) in reef organisms living directly seaward of river mouths may have reached the Great Barrier Reef on sediment flocs. Studies of cohesive sediment transport in the Great Barrier Reef are thus warranted.

Long-term studies, including more sensitive methods for measuring the impacts of dredging and spoil dumping, as well a studies on the long-term fate of dredge spoil dumped on the continental shelf of the Great Barrier Reef, are needed in these critical environments where some changes may be irreversible (Figure 10.1).

Studies are also urgently needed to investigate the nature and fate of farm effluents in the coastal environment with a view to recommending safe levels and/or alternative disposal methods. Such recommendations need to be based on a sound knowledge of the dynamics of cohesive sediments advected by a baroclinic circulation around a rugged topography. This knowledge is presently unavailable, nor is there at present research undertaken in that area.

Man has already significantly contributed to the degradation of the system. Major problems caused by ever-increasing human pressure on the Great Barrier Reef will not go away, but instead will keep growing. Multidisciplinary, process-oriented research based on a sound, reliable understanding of oceanographic processes is needed to help provide answers to key questions necessary to wisely manage the Great Barrier Reef. Without this research, the future of the Great Barrier Reef will be at increased risk because some management decisions will be based on ill-founded or nonconservative assumptions.

References

Alldredge, A. L. and W. H. Hammer (1980). Recurring aggredation of zooplankton by a tidal current. *Estuarine and Coastal Marine Science* 10, 31–37.

Alongi, D. M. (1990). Effects of mangrove detrital outwelling on nutrient regeneration and oxygen fluxes in coastal sediments of the central Great Barrier Reef lagoon. *Estuarine, Coastal and Shelf Science* 31, 581–598.

Amin, M. (1978). A statistical analysis of storm surges in Torres Strait. *Australian Journal of Marine and Freshwater Research* 29, 479–496.

Andrews, J. C. (1983a). Thermal waves on the Queensland shelf. *Australian Journal of Marine and Freshwater Research* 34, 81–96.

Andrews, J. C. (1983b). Water masses, nutrient levels and seasonal drift on the outer central Queensland shelf (Great Barrier Reef). *Australian Journal of Marine and Freshwater Research* 34, 821–834.

Andrews, J. C. and P. Gentien (1982). Upwelling as a source of nutrients for the Great Barrier Reef ecosystems: a solution to Darwin's question. *Marine Ecology Progress Series* 8, 257–269.

Andrews, J. C. and H. Muller (1983). Space-time variability of nutrients in a lagoonal patch reef. *Limnology and Oceanography* 28, 215–227.

Andrews, J. C. and M. Furnas (1986). Subsurface intrusions of Coral Sea water into the central Great Barrier reef. I. Structures and shelf-scale dynamics. *Continental Shelf Research* 6, 491–514.

Andrews, J. C. and L. Bode (1988). The tides of the central Great Barrier Reef. *Continental Shelf Research* 8, 1057–1085.

Andrews, J. C. and S. Clegg (1989). Coral sea circulation deduced from modal information models. *Deep-Sea Research* 36, 957–974.

Andrews, J. C., A. Mitchell and N. F. Bellamy (1983). Field studies of currents and simulation of oil dispersion in the central Great Barrier Reef. *Proceedings 6th Australasian Conference Coastal and Ocean Engineering,* Brisbane, 83/6, 80–84.

Apelt, C. J. and N. J. Richter (1985). Modelling Barrier Reef tides. *Trans. Institute of Engineers, Australia,* 27, 166–173.

Atkinson, M. J. and R. W. Bilger (1992). Effects of water velocity on phosphate uptake in coral reef flat communities. *Limnology and Oceanography* 37, 273–279.

Awaji, T. A. (1982). Water mixing in a tidal current and the effect of turbulence on tidal exchange through a strait. *Journal of Physical Oceanography* 12, 501–514.

Awaji, T. A., N. Asimoto and K. Kunishi (1980). Tidal exchange through a strait: a numerical experiment using a simple model basin. *Journal of Physical Oceanography* 10, 1499–1508.

Baines, P. G. (1982). On internal tide generation models. *Deep-Sea Research* 29, 307–338.

Baines, P. G. and P. A. Davies (1980). Laboratory studies of topographic effects in rotating and/ or stratified fluids. In: R. Hide and P. White (eds.) , *Orographic Effects in Planetary Flows.* GARP Publishing Service 23, WMO, Geneva, 223–299.

Bakun, A. (1986). Local retention of planktonic early life stages in tropical reef bank demersal systems: the role of vertically-structured hydrodynamic processes. IOC/FAO Workshop on Recruitment in Tropical Coastal Demersal Communities, Campeche, 21–25 April 1986, 15–32.

Barstow, S. F. (1983). The ecology of Langmuir circulation: a review. *Marine Environmental Research* 9, 211–236.

Batchelor, G. K. (1967). *An Introduction to Fluid Dynamics*. Cambridge University Press, Cambridge, 615 pp.

Battisti, D. S. and A. J. Clarke (1982a). A simple method for establishing barotropic tidal currents on continental margins with specific application to M2 tide off the Atlantic and Pacific coasts of the United States. *Journal of Physical Oceanography* 12, 8–16.

Battisti, D. S. and A. J. Clarke (1982b). Estimation of near-shore tidal currents on nonsmooth continental shelves. *Journal of Geophysical Research* 87, 7873–7878.

Bell, P. R. F. (1990). Impact of run-off and wastewater discharges on eutrophication in the Great Barrier Reef region. *Proceedings Conference on Engineering in Coral Reef Regions*, Townsville, 57 November, 1990, 111–115.

Bellamy, N., A. Mitchell, P. Gentien, J. C. Andrews and S. Ball (1982). Oceanographic observations on the outer slope and slope in the central zone of the Great Barrier Reef. Australian Institute of Marine Science, Data Report, AIMS-OS-82-2.

Belperio, A. P. (1979). The combined use of wash load and bed material load rating curves for the calculation of total load: an example from the Burdekin River, Australia. *Catena* 6, 317–329.

Belperio, A. P. (1983). Terrigenous sedimentation in the central Great Barrier reef lagoon: a model from the Burdekin region. *BMR Journal of Geology and Geophysics* 8, 179–190.

Benson, A. A. and L. Muscatine (1974). Wax in coral mucus: energy transfer from corals to reef fishes. *Limnology and Oceanography* 19, 810–814.

Black, K. P. and S. L. Gay (1987). Eddy formation in unsteady flows. *Journal of Geophysical Research* 92, 9514–9522.

Black, K. P. and P. J. Moran (1991). Influence of hydrodynamics on the passive dispersal and initial recruitment of larvae of *Acanthaster planci* (Echinodermata: Asteroidea) on the Great Barrier Reef. *Marine Ecology Progress Series* 69, 55–65.

Black, K. P., S. L. Gay and J. C. Andrews (1990). Residence times of neutrally-buoyant matter such as larvae, sewage or nutrients on coral reefs. *Coral Reefs* 9, 105–114.

Black, K. P., P. J. Moran and L. S. Hammond (1991). Hydrodynamic models show coral reefs can be self-seeding. *Marine Ecology Progress Series* 74, 1–11.

Blumberg, A. F. and G. L. Mellor (1987). A description of a three-dimensional coastal ocean circulation model. In: N. S. Heaps (ed.), *Three-dimensional Coastal Ocean Models*. American Geophysical Union, Washington, D.C., 1–16.

Bode, L. and R. J. Sobey (1983). The open boundary problem for wind-driven circulations. *Proceedings 8th Australasian Conference on Fluid Mechanics*, Newcastle, 4A1–4A4.

Bode, L. and M. Stark (1983). Simulation of tides and currents in the Mackay region. *Proceedings 6th Australasian Conference on Coastal and Ocean Engineering, Brisbane*, 83/6, 85–89.

Bode, L., L. B. Mason, R. J. Sobey and K. P. Stark (1981). Hydrodynamic studies of water movements within the Great Barrier Reef region. I. Preliminary investigations. Department of Civil and Systems Engineering, James Cook University, Research Bulletin CS-27, 125 pp.

Boehlert, G. W., W. Stanton and L. C. Sun (1992). Horizontal and vertical distributions of larval fishes around an isolated oceanic island in the tropical Pacific. *Deep-Sea Research* 39, 3/4, 439–466.

Boto, K. and P. Isdale (1985). Fluorescent bands in massive corals result from terrestrial fulvic acid inputs to nearshore zone. *Nature* 315, 396–397.

Boto, K. G. and A. I. Robertson (1990). The relationship between nitrogen fixation and tidal exports of nitrogen in a tropical mangrove system. *Estuarine, Coastal and Shelf Science* 31, 531–540.

Boyer, D. L. and P. A. Davies (1982). Flow past a circular cylinder on a β-plane. *Philosophical Transactions of the Royal Society of London Ser.* A,306, 533–556.

Breitburg, D. L. (1989). Demersal schooling prior to settlement by larvae of the naked goby. *Environmental Biology of Fishes* 26, 97–103.

Brink, K. H. and J. S. Allen (1978). Models of wind-driven currents on the continental shelf. *Annual Review of Fluid Mechanics* 12, 389–433.

Buchwald, V. T. and J. W. Miles (1981). On resonance of an offshore channel bounded by a reef. *Australian Journal of Marine and Freshwater Research* 32, 833–841.

Burgraff, O. R., K. Stewartson and R. Belcher (1971). Boundary layer induced by a potential vortex. *Physics of Fluids* 14, 1821–1833.

Burrage, D. M., J. A. Church and C. R. Steinberg (1991). Linear systems analysis of momentum on the continental shelf and slope of the central Great Barrier Reef. *Journal of Geophysical Research* 96, C12, 22169–22190.

Cameron McNamara, Consulting Engineers (1985). Stream aggradation, effects of soil erosion. Cameron McNamara Report 84-1501, 54 pp.

Church, J. A. (1983). A review of the circulation in the Western Coral Sea and internal tides on the shelf slope adjacent to the Great Barrier Reef. *Proceedings I Great Barrier Reef Conference,* Townsville, pp. 415–420.

Church, J. A. and F. M. Boland (1984). A permanent undercurrent adjacent to the Great Barrier Reef. *Journal of Physical Oceanography* 13, 1747–1749.

Church, J. A.., J. C. Andrews and F. M. Boland (1985). Tidal currents in the central Great Barrier Reef. *Continental Shelf Research* 4, 515–531.

Collins, J. D. and T. A. Walker (1985). A drift card study of the Great Barrier Reef. Final Report to Great Barrier Reef Marine Park Authority.

Colman, R. S., H. C. Crenshaw, D. L. Meyer and J. R. Strickler (1984). A non-motorised dye ejector for visualization of flow in situ and its use with coral reef crinoids. *Marine Biology* 83, 125–128.

Craik, W. (1990). Oil spills in the Great Barrier Reef region. *Proceedings Conference on Engineering in Coral Reef Regions,* Townsville, 57 November, 1990, 211–222.

Csanady, G. T. (1976). Mean circulation in shallow seas. *Journal of Geophysical Research* 81, 5389–5399.

Csanady, G. T. (1978). The arrested topographic wave. *Journal of Physical Oceanography* 8, 47–62.

Csanady, G. T. (1982). *Circulation in the coastal ocean.* Reidel Publishing Co., Sordrecht.

Davies, P. J. and B. G. West (1981). Suspended sediment and water movement at One Tree Reef, southern Great Barrier Reef. *BMR Journal of Australian Geology and Geophysics* 6, 187–195.

Davies, P. J. and H. Hughes (1983). High-energy reef and terrigenous sedimentation. Boulder Reef, Great Barrier Reef. *BMR Journal of Australian Geology and Geophysics* 8, 201–209.

Davis, C. S., S. M. Gallagher and A. R. Solow (1992). Microaggregations of oceanic plankton observed by towed video microscopy. *Science* 257, 230–232.

Deacon, E. L. (1979). The role of coral mucus in reducing the wind drag over coral reefs. *Boundary-Layer Meteorology* 17, 517–521.

Deleersnijder, E., A. Norrow and E. Wolanski (1992). A three-dimensional model of the water circulation around an island in shallow water. *Continental Shelf Research* 12, 891–906.

De Vaugelas, J. (1985). Sediment reworkings by Callianassid mud-shrimp in tropical lagoons: a review with perspectives. *Proceedings 5th International Coral Reef Congress* 6, 617–622.

Dight, I. J., L. Bode and M. K. James (1990a). Modeling the larval dispersal of *Acanthaster planci*. I. Large scale hydrodynamics, Cairns section, Great Barrier Reef marine park. *Coral Reefs* 9, 115–123.

Dight, I. J., M. K. James and L. Bode (1990b). Modeling the dispersal of *Acanthaster planci*. *Coral Reefs* 9, 125–134.

Done, T. J. (1992). Effects of tropical cyclones on ecological and geomorphological structures on the Great Barrier Reef. *Continental Shelf Research* 12, 859–872.

Downey, W. K. (1983). Meteorology of the Great Barrier Reef and western Coral Sea. *Proceedings International Great Barrier Reef Conference*, Townsville, 421–433.

Dufour, V. (1992). Colonisation des recifs coralliens par les larves de poissons. These de l'Universite Pierre et Marie Curie, Paris, 220 pp.

Easton, A. K. (1970). The tides of the continent of Australia. Research Paper No. 37. Horace Lamb Centre for Oceanographical Research, Flinders University of South Australia, 326 pp.

Falconer, R. A., E. Wolanski and L. Mardapitta-Hadjipandeli (1986a). Modeling tidal circulation in an island's wake. American Society of Civil Engineers, *J. Waterway, Port, Coastal & Ocean Engineering* 112, 2, 234–254.

Falconer, R. A., E. Wolanski and L. Mardapitta-Hadjipandeli (1986b). Integrated field measurements and numerical model simulations of tidal eddies. *Proceedings International Conference on Measuring Techniques of Hydraulics Phenomena in Offshore, Coastal and Inland Waters*, London, 9–11 April, 1986.

Fischer, H. B., E. Y. List, R. C. Y. Koh, J. Imberger and N. H. Brooks (1979). *Mixing in Inland and Coastal Waters*. Academic Press, New York, 483 pp.

Foreman, M. G. G. (1977). Manual for tidal height analysis and prediction. Pacific Marine Science Rep. 77-10. Institute of Ocean Science, Patricia Bay, Sidney, British Columbia, 97 pp.

Franks, P. J. S. (1992). Phytoplankton blooms at fronts: patterns, scales, and physical forcing mechanisms. *Reviews in Aquatic Sciences*, CRC Press, 6, 2, 121–137.

Fu, L. L. and B. Holt (1982). Seasat views oceans and sea ice with synthetic aperture radar. Publication 81-120. Jet Propulsion Laboratory, California Institute of Technology, Pasadena, California.

Furnas, M. J. (1989). Cyclonic disturbance and a phytoplankton bloom in a tropical shelf ecosystem. In: T. Okaichi, D. M. Anderson and T. Nemoto (eds.), *Red Tides: Biology, Environmental Science, and Toxicology*, Elsevier Science Publishing Co., New York, pp. 273–276.

Furnas, M. J. (1992). Pelagic Trichodesmium (= Oscillatoria) in the Great Barrier Reef region. In: E. J. Carpenter (ed.), *Marine Pelagic Cyanobacteria: Trichodesmium and Other Diazotrophs*. Kluwer Academic Press, The Netherlands, 265–272.

Furnas, M. and A. W. Mitchell (1986). Phytoplankton dynamics in the central Great Barrier Reef. I. Seasonal changes in biomass and community structure and their relation to intrusive activity. *Continental Shelf Research* 6, 3, 363–384.

Furnas, M. and A. W. Mitchell (1987). Phytoplankton dynamics in the central Great Barrier Reef. II. Primary production. *Continental Shelf Research* 7, 9, 1049–1062.

Gabric, A. J., P. Hoffenerg and W. Boughton (1990). Spatio-temporal variability in surface chlorophyll distribution in the central Great Barrier Reef as derived from CZCS imagery. *Australian Journal of Marine and Freshwater Research* 41, 313–324.

Gagan, M. K., M. W. Sandstrom and A. R. Chivas (1987). Restricted terrestrial carbon input to the continental shelf during cyclone Winifred: implication for terrestrial runoff to the Great Barrier Reef province. *Coral Reefs* 6,, 113–119.

Gagan, M. K., D. P. Johnson and R. M. Carter (1988). The cyclone Winifred storm bed, central Great Barrier Reef shelf, Australia. *Journal of Sedimentary Petrology* 58, 845–856.

Gay, S. L., J. C. Andrews and K. P. Black (1990). Dispersal of neutrally buoyant material near John Brewer Reef. In: R. Bradbury (ed.), *Acanthaster in the Coral Reef: A Theoretical Perspective*. Lecture notes in Biomathematics, Springer-Verlag, Berlin, 95–199.

Gerrard, T. H. (178). The wake of cylindrical bluff bodies at low reynolds number. *Philosophical Transactions of the Royal Society of London* 288, 351–382.

Gerritsen, F. (1981). Wave attenuation and set-up on a coastal reef. Look Laboratory Technical Report No. 48, University of Hawaii, 416 pp.

Geyer, W. R. (1993). Three-dimensional tidal flow around headlands. *Journal of Geophysical Research,* 98, 955–966.

Geyer, W. R. and R. Signell (1900). Measurements of tidal flow around a headland with a shipboard acoustic Doppler current profiler. *Journal of Geophysical Research* 95, 3189–3197.

Gibbs, R. J. (1993). Metals in the bottom muds in Townsville harbour, Australia. *Environmental Pollution,* 81, 297–300.

Gidhagen, L. (1984). Coastal upwelling in the Baltic. Swedish Meteorological and Hydrological Institute, Norrkoping.

Gill, A. E. and E. H. Schumann (1974). The generation of long shelf waves by the wind. *Journal of Physical Oceanography* 4, 83–90.

Gordon, H. B. and R. L. Hughes (1981). A study of rotating baroclinic nonlinear flow around an island. *Journal of Physical Oceanography* 11, 1011–1014.

Gourlay, M. R. (1988). Coral cays: products of wave action and geological processes in a biogenic environment. *Proceedings 6th International Coral Reef Symposium,* Australia, 1988, 2, 491–496.

Gourlay, M. R. and C. J. McMonagle (1989). Cyclonic wave prediction in the Capricornia region, Great Barrier Reef. *Proceedings 9th Australasian Conference on Coastal and Ocean Engineering,* Adelaide, 48 December, 1989, 112–116.

Graham, H. E. and G. N. Hudson (1960). Surface winds near the centre of hurricanes (and other cyclones). National Hurricane Research Project, Report No. 39. U.S. Department of Commerce.

Grant, W. D. and O. S. Marsden (1979). Combined wave and current interaction with a rough bottom. *Journal of Geophysical Research* 84, 1797–1808.

Grant, W. D., A. J. Williams and S. M. Glenn (1984). Bottom stress estimates and their prediction in the northern California continental shelf during CODE-1: the importance of wave-current interaction. *Journal of Physical Oceanography* 14, 506–527.

Greenspan, H. P. (1968). *The Theory of Rotating Fluids.* Cambridge University Press, New York.

Griffin, D. A. and J. H. Middleton (1986). Coastal-trapped waves behind a large continental shelf island, southern Great Barrier Reef. *Journal of Physical Oceanography* 16, 1651–1664.

Griffin, D. A., J. H. Middleton and L. Bode (1987). The tidal and longer period circulation of Capricornia, southern Great Barrier Reef, *Australian Journal of Marine and Freshwater Research* 38, 461–474.

Hamner, W. M. and I. R. Hauri (1977). Fine-scale currents in the Whitsunday Islands, Queensland, Australia: effect of tide and topography, *Australian Journal of Marine and Freshwater Research* 28, 333–359.

Hamner, W. M. and I. Hauri (1981). Effects of island mass: water flow and plankton pattern around a reef in the Great Barrier Reef lagoon, Australia. *Limnology and Oceanography* 26, 1084–1102.

Hamner, W. M. and D. Schneider (1986). Regularly spaced rows of medusae in the Bering Sea: role of Langmuir circulation. *Limnology and Oceanography* 31, 171–177.

Hamner, W. M., M. S. Jones, J. H. Carleton, I. R. Hauri and D. McB. Williams (1988). Zooplankton, planktivorous fish, and water current on a windward reef face: Great Barrier reef, Australia. *Bulletin of Marine Science* 42, 3, 459–479.

Hamon, B. V. (1984). Diurnal and semi-diurnal tides on the Great Barrier Reef. CSIRO Marine Laboratories (Australia) Report 163, 20 pp.

Hardy, T. A., I. R. Young, R. C. Nelson and M. R. Gourlay (1990). Wave attenuation on an offshore coral reef. *Proceedings 22nd International Coastal Engineering Conference*, Delft, 26 July, 1990, Vol. 1, pp. 330–344.

Hart, B. T., G. Day, A. Sharp-Paul and T. Beer (1988). Water quality variations during a flood event in the Annan River, North Queensland. *Australian Journal of Marine and Freshwater Research* 39, 225–243.

Heywood, K. J., E. D. Barton and J. H. Simpson (1990). The effects of flow disturbance by an oceanic island. *Journal of Marine Research* 48, 55–73.

Hicks, B. B., R. L. Brinkrow and G. Krause (1974). Drag and bulk transfer coefficients associated with a shallow water surface. *Boundary-Layer Meteorology* 6, 287–297.

Hinwood, J. B., D. R. Blackman and G. T. Leonart (1982). Some properties of swell in the southern ocean. *Proceedings 18th International Conference Coastal Engineering*, ASCE, Cape Town, 261–269.

Hogg, N. G. (1972). Steady flow past an island with applications to Bermuda. *Geophysical Fluid Dynamics* 4, 55–81.

Hogg, N. G. (1980). Effects of bottom topography on ocean currents. In: R. Hide and P. White (eds.). Orographic effects in planetary flows, GARP Publications Ser. 23, W.M.O., Geneva, 167–205.

Hogg, N. G., E. Katz and T. Stanford (1978). Eddies, islands, and mixing, *Journal of Geophysical Research* 83, 2921–2938.

Hopley, D. (1982). *The Geomorphology of the Great Barrier Reef*. Wiley, New York, 453 pp.

Hopley, D. (1990). The geology and geomorphology of the Great Barrier Reef in relation to engineering problems. *Proceedings Conference on Engineering in Coral Reef Regions*, Townsville, 57 November, 1990, 61–74.

Horikawa, K. and C. T. Kuo (1967). A study of wave transformation inside the surf zone. *Proceedings 10th International Conference on Coastal Engineering*, ASCE, I, 217–233.

Huang, N. E., S. R. Long, C. C. Tong, Y. Yuen and L. F. Bliuen (1981). A unified two-parameter wave spectral model for a general sea state. *Journal of Fluid Mechanics* 112, 203–204.

Huthnance, J. M. (1985). Flow across reefs or between islands and effects on shelf-sea motions. *Continental Shelf Research* 4, 709–731.

Ikeda, M. and W. J. Emery (1984). Satellite observations and modeling of meanders in the California current system off Oregon and northern California. *Journal of Physical Oceanography* 14, 1434–1450.

Isdale, P. (1984). Fluorescent bands in massive corals record centuries of coastal rainfall. *Nature* 310, 578–579.

Johannes, R. E. (1967). Ecology of organic aggregates in the vicinity of a coral reef. *Limnology and Oceanography* 12, 189–195.

Johnson, D. P. and R. M. Carter (1987). Nature of dredge spoil taken by the Townsville harbour board. Unpublished manuscript, Department of Geology, James Cook University.

Keller, H. B. and F. Niewstadt (1973). Viscous flow past circular cylinders. *Computers and Fluids* 1, 59–71.

Kennedy, V. S. (ed.) (1984). *The Estuary as a Filter*. Academic Press, New York, 511 pp.

King, B. (1992). A predictive model in the currents in Cleveland Bay. In: M. Spaulding, K. Bedford, A. Blumberg, R. Cheng and C. Swanson (eds.), *Estuarine and Coastal Modeling.* American Society of Civil Engineers, New York, 786 pp.

King, B. and E. Wolanski (1990). Coastal dynamics along a rugged coastline. In: D. Prandle (ed.), *Dynamics and Exchanges in Estuaries and the Coastal Zone.* A.G.U. Coastal and Estuarine Series, New York, 647 pp.

Kingsford, M. J., E. Wolanski and J. H. Choat (1991). Influence of tidally induced fronts and Langmuir circulations on distribution and movements of presettlement fishes around a coral reef. *Marine Biology* 109, 167–180.

Kjerfve, B. and S. P. Dinnel (1983). Hindcasting hurricane characteristics on the Belize Barrier Reef. *Coral Reefs* 1, 203–207.

Kjerfve, B., K. E. Macgill, J. W. Porter and J. D. Woodley (1986). Hindcasting of hurricane characteristics and observed storm damage on a fringing reef, Jamacia, West Indies. *Journal of Marine Research* 44, 119–148.

Lee, T. T. and K. P. Black (1979). The energy spectra of surf waves on a coral reef. *Proceedings 16th International Conference on Coastal Engineering,* ASCE, 1, 588–608.

Liston, P., M. J. Furnas, A. W. Mitchell and E. A. Drew (1992). Local and mesoscale variability of surface water temperature and chlorophyll in the northern Great Barrier Reef, Australia. *Continental Shelf Research* 12, 907–922.

Marshall, J. F. (1985). Cross-shelf and facies related variations in submarine cementation in the central Great Barrier Reef. *Proceedings 5th International Coral Reef Symposium* 3, 221–226.

Massel, S. R. (1989). *Hydrodynamics of Coastal Zones.* Elsevier Scientific Publishing Co., 336 pp.

Massel, S. R. (1992). Wave transformation and dissipation on steep reef slopes. Proceedings 11th Australasian Fluid Mechanics Conference, Hobart, 1992, 319–322.

Middleton, J. H. (1983). Low-frequency trapped waves on a wide, reef-fringed continental shelf. *Journal of Physical Oceanography* 13, 1371–1382.

Middleton, J. H. (1987). Steady coastal circulation due to oceanic alongshore pressure gradients. *Journal of Physical Oceanography* 17, 604–612.

Middleton, J. H. and A. Cunningham (1984). Wind-forced continental shelf waves from a geographical origin. *Continental Shelf Research* 3, 215–232.

Middleton, J. A. H., V. T. Buchwald and J. M. Huthnance (1984). The anomolous tides near Broad Sound. *Continental Shelf Research* 3, 359–381.

Middleton, J. A. H., D. A. Griffin and A. M. Moore (1993). Oceanic circulation and turbulence in the coastal zone. *Continental Shelf Research* 13, 143–168.

Millot, C. (1979). Wind-induced upwellings in the Gulf of Lions. *Oceanologica Acta* 2, 261–274.

Mulhearn, P. J. (1989). Turbidity in Torres Strait. Defence Science and Technology Organisation, Technical Memorandum WSRL-TM-35/89, 29 pp.

Munk, W. M. and M. S. Sargent (1954). Adjustment of Bikini Atoll to ocean waves. *U.S. Geological Survey Prof. Papers* 260-C, 275–280.

Murray, R. T. and L. R. Ford (1983). Problems in the analysis of data for the assessment of longshore sediment transport: an example from North Queensland, *Proceedings 6th Australasian Conference on Coastal and Ocean Engineering,* Institute of Engineers, Australia.

Murray, S. P. and M. Young (1985). The nearshore current along a high rainfall tradewind coast — Nicaragua. *Estuarine, Coastal and Shelf Science* 21, 687–699.

Nelson, R. C. and E. J. Lesleighter (1985). Breaker height attenuation over platform coral reefs. *Proceedings 7th Australasian Conference on Coastal and Ocean Engineering, Institute* of Engineers, Australia, 2, 9–16.

Nihoul, J. C. J., E. Deleersnijder and S. Djenidi (1989). Modelling the general circulation of shelf seas by 3D k-e models. *Earth-Science Reviews* 26, 163–189.

Nof, D. (1981). On the dynamics of equatorial outflows with applications to the Amazon Basin. *Journal of Marine Research* 39, 129.

Nof, D. and J. H. Middleton (1989). Geostrophic pumping, inflows and upwelling in barrier reefs. *Journal of Physical Oceanography* 19, 7, 874–889.

Oberdorfer, J. A. and R. W. Buddemeier (1986). Coral reef hydrology: field studies of water movement within a reef barrier. *Coral Reefs* 5, 7–12.

Okubo, A. (1973). Effect of shoreline irregularities on steamwise dispersion in estuaries and other embayments. *Netherlands Journal of Sea Research* 6, 213–224.

Okubo, A. (1974). Some speculations on oceanic diffusion diagrams. *Rapp. P.-V. Reunion Con. Int. Explor. Mer* 167, 77–85.

Olafson, R. W. (1978). Effects of agricultural activity on levels of organochlorine pesticides in hard corals, fish and molluscs from the Great Barrier Reef. *Marine Environmental Research* 1, 87–107.

Oliver, J. K. and B. L. Willis (1987). Coral-spawn slicks in the Great Barrier Reef: preliminary observations. *Marine Biology* 94, 521–529.

Oliver, J. K., B. A. King, B. L. Willis, R. C. Babcock and E. Wolanski (1992). Dispersal of coral larvae from a lagoonal reef. II. Comparisons between model predictions and observed concentrations. *Continental Shelf Research* 12, 873–889.

Onishi, S. (1984). Study of vortex structure in water surface jets by means of remote sensing. In: J. C. J. Nihoul (ed.), *Remote Sensing of Shelf Seas Hydrodynamics*. Elsevier, Amsterdam, pp. 107–132.

Onishi, S. (1986). Roles of large-scale eddies in mass exchange between coastal and oceanic regions. In: J. van de Kreeke (ed.), *Physics of Shallow Estuaries and Bays*. Springer-Verlag, New York, 168–177.

Parnell, K. (1988). Physical process studies in the Great Barrier Reef Marine Park. *Progress in Physical Geography* 12, 209–236.

Patterson, M. R., K. P. Sebens and R. R. Olson (1991). *In situ* measurements of flow effects on primary production and dark respiration in reef corals. *Limnology and Oceanography* 36, 936–948.

Pattiaratchi, C., A. James and M. Collins (1987). Island wakes and headland eddies: a comparison between remotely sensed data and laboratory experiments. *Journal of Geophysical Research* 92, 783–794.

Phillips, O. M. J., J. Shyu and H. Salmun (1986). An experiment on boundary mixing: mean circulation and transport rates. *Journal of Fluid Mechanics* 173, 473–499.

Pickard, G. L. (1986). Effects or wind and tide on upper layer currents at Davies Reef, Great Barrier Reef, during MECOR (July/August 1984). *Australian Journal of Marine and Freshwater Research* 37, 545–565.

Pickard, G. L., J. R. Donguy, C. Henin and F. Rougerie (1977). A review of the physical oceanography of the Great Barrier Reef and western Coral Sea. Australian Institute of Marine Science, Monograph Series Vol. 2, 134 pp.

Piorewicz, J. and B. J. Saunders (1989). Changes to Yeppoon Main beach— the engineering process (a case study). Proceedings Central Queensland Engineering Conference, Rockhampton, 1517 September, 1989, C5, 1–11.

Piorewicz, J. and G. L. Ryall (1991). Hydrosedimentological studies of Johnstone Estuary (North Queensland). *Proceedings 10th Australasian Conference on Coastal and Ocean Engineering*, Auckland, 26, December, 1991, 459–464.

Pollard, R. T. (1976). Observations and theories of Langmuir circulations and their role in near surface mixing. *Deep-Sea Research, Sir George Deacon Anniversary Supplement* 235–251.

Prager, E. J. (1991). Numerical simulation of circulation in a Caribbean-type backreef lagoon. *Coral Reefs* 10, 177–182.

Pringle, A. W. (1989). The history of dredging in Cleveland Bay, Queensland, and its effect on sediment movement and on the growth of mangroves, corals and seagrass. Great Barrier Reef Marine Park Authority Research Publication, 177 pp.

Provis, D. G. and G. W. Lennon (1983). Eddy viscosity and tidal cycles in a shallow sea. *Estuarine, Coastal and Shelf Science* 31, 541–555.

Ratcliffe, F. (1947). *Flying Fox and Drifting Sand.* Angus & Robertson, Melbourne, 332 pp.

Ridd, P., E. Wolanski and Y. Mazda (1990). Longitudinal diffusion in mangrove-fringed tidal creeks. *Estuarine, Coastal and Shelf Science* 31, 541–555.

Roberts, H. H. H. (1981). Physical processes and sediment flux through reef-lagoon systems. *Proceedings 17th International Conference on Coastal Engineering,* ASCE, 1, 946–962.

Roberts, H. H. H., S. P. Murray and J. N. Suhayada (1975). Physical processes in a fringing reef system. *Journal of Marine Research* 33, 233–260.

Roberts, H. H. H., S. P. Murray and J. N. Suhayada (1977). Physical processes in a fore-reef shelf environment. *Proceedings 3rd International Coral Reef Symposium,* Miami, 2, 508–515.

Roberts, H. H. H., W. J. Wiseman and T. H. Suchanek (1981). Lagoon sediment transport: the significant effect of *Callianassa* bioturbation. *Proceedings 4th International Coral Reef Symposium* 1, 459–465.

Roberts, H. H., A. Lugo, B. Carter and M. M. Simms (1988). Across-reef flux and shallow subsurface hydrology in modern reef corals. *Proceedings 6th International Coral Reef Symposium* 2, 509–515.

Roberts, H. H. H., P. A. Wilson and A. Lugo-Fernandez (1992). Biologic and geologic responses to physical processes: examples from modern reef systems of the Caribbean-Atlantic region. *Continental Shelf Research* 12, 809–834.

Robertson, A. I., D. M. Alongi, P. Daniel and K. G. Boto (1988). How much mangrove detritus enters the Great Barrier Reef lagoon? *Proceedings 6th International Symposium on Coral Reefs,* Townsville, August 1988.

Ross, M. A. and A. J. Mehta (1989). On the mechanics of lutoclines and fluid mud. *Journal of Coastal Research* 5, 51–61.

Sahl, L. E. and M. A. Marsden (1987). Shelf sediment dispersal during the dry season, Princess Charlotte Bay, Great Barrier Reef, Australia. *Continental Shelf Research* 7, 10, 1139–1159.

Scandol, J. P. and M. K. James (1992). Hydrodynamics and larval dispersal: a population model of *Acanthaster planci* on the Great Barrier Reef. *Australian Journal of Marine and Freshwater Research* 43, 583–596.

Schumann, E. H., L. A. Perrins and I. T. Hunter (1982). Upwelling along the south coast of Cape Province, South Africa. *South African Journal of Science* 78, 238–242.

Sebens, K. P. and A. S. Johnson (1991). Effects of water movement on prey capture and distribution of reef corals. *Hydrobiologia* 226, 91–101.

Shanks, A. L. (1983). Surface slicks associated with tidally forced internal waves may transport pelagic larvae of benthic invertebrates and fishes shoreward. *Marine Ecology Progress Series* 13, 311–315.

Signell, R. P. (1989). Tidal dynamics and dispersion around coastal headlands. Ph.D. thesis, MIT/WHOI-89-38.

Signell, R. P. and W. R. Geyer (1990). Numerical simulation of tidal dispersion around a coastal headland. In: R. T. Cheng (ed.). *Residual Currents and Long-Term Transport in Estuaries and Bays*. Lecture notes on Coastal and Estuarine Studies, Springer-Verlag, 210–222.

Signell, R. P. and W. R. Geyer (1991). Transient eddy formation around headlands. *Journal of Geophysical Research* 96, 2561–2575.

Signell, R. P., R. C. Beardsley, H. C. Graber and A. Capotondi (1990). Effect of wave-current interaction on wind-driven circulation in narrow, shallow embayments. *Journal of Geophysical Research* 95, 9671–9678.

Simpson, J. H. and I. D. James (1986). Convergent fronts in the circulation of tidal estuaries. In: D. A. Wolfe (ed.), *Estuarine Variability. Academic* Press, London.

Simpson, J. H., P. B. Tett, M. L. Argote-Espinoza, A. Edwards, K. J. Jones and G. Savidge (1982). Mixing and phytoplankton growth around an island in a stratified sea. *Continental Shelf Research* 1, 1, 15–31.

Snodgrass, F. E., G. W. Groves, K. F. Hasselman, G. R. Miller, W. H. Munk and W. M. Munk (1966). Propagation of ocean swell across the Pacific. *Philosophical Transactions of the Royal Society of London, Ser. A*, 259, 431–497.

Sobey, R. J., B. A. Harper and G. M. Mitchell (1982). Numerical modelling of tropical cyclone storm surges. Civil Engineering Trans., Institute of Engineers, Australia, 24, 151–161.

Stern, M. E. (1980). Geostrophic fronts, bores, braking and blocking waves. *Journal of Fluid Mechanics* 99, 4, 687–703.

Takahashi, M., Y. Yasuoka, M. Matanabe, T. Miyazaki and S. Ichimura (1981). Local upwelling associated with vortex motion off Oshima Island. In: F. Richards (ed.), *Coastal Upwelling*. A.G.U., Washington, D.C.

Talbot, M. M. B. and G. C. Bate (1987). Rip current characteristics and their role in the exchange of water and turf diatoms between the surf zone and nearshore. *Estuarine, Coastal and Shelf Science* 25, 707–720.

Thompson, R. O. R. Y. and T. J. Golding (1981). Tidally induced upwelling by the Great Barrier Reef. *Journal of Geophysical Research* 86, 6517–6521.

Thomson, R. E. and E. Wolanski (1984). Tidal period upwelling within Raine Island Entrance, Great Barrier Reef. *Journal of Marine Research* 42, 787–808.

Tomczak, M. (1983). Coral Sea water masses. *Proceedings International Great Barrier Reef Conference*, Townsville, 461–466.

Tomczak, M. (1988). Island wakes in deep and shallow water. *Journal of Geophysical Research* 89, 10553–10569.

Tomczak, M. and X. H. Fang (1983). Attempts to determine some properties of the semi-diurnal tide on the continental slope. Great Barrier Reef. *Australian Journal of Marine and Freshwater Research*, 34, 6, 921–926.

Tudhope, A. W. and T. P. Scoffin (1984). The effects of *Callianassa* bioturbation on the preservation of carbonate grains in Davies Reef lagoon, Great Barrier Reef, Australia. *Journal of Sedimentary Petrology* 54, 1091–1096.

van Senden, D. C. and J. Imberger (1990). Effects of initial conditions and Ekman suction on tidal outflows from inlets. *Journal of Geophysical Research* 95, 13373–13391.

Victor, B. C. (1984). Coral reef fish larvae: patch size estimation and mixing in the plankton. *Limnology and Oceanography* 29, 1116–1119.

von Riegels, F. (1938). Zur kritik des Hele-Shaw Versuchs. *Z. Angew. Math. Mech.* 18, 95–106.

Walker, T. A. (1981a). Seasonal salinity variations in Cleveland Bay, northern Queensland. *Australian Journal of Marine and Freshwater Research* 32, 143–149.

Walker, T. A. (1981b). Annual temperature cycle in Cleveland Bay, Great Barrier Reef Province. *Australian Journal of Marine and Freshwater Research* 32, 987–981.

Walker, T. A. (1982). Lack of Evidence for evaporation-driven circulation in the Great Barrier Reef lagoon. *Australian Journal of Marine and Freshwater Research* 33, 717–722.

Walker, T. A. and G. O'Donnell (1981). Observations of nitrate, phosphate and silicate in Cleveland Bay, northern Queensland. *Australian Journal of Marine and Freshwater Research* 877–887.

Walker, G. R., G. F. Reardon and E. D. Jancauskas (1988). Observed effects of topography on the wind field of cyclone Winifred. *Journal of Wind Engineering and Industrial Aerodynamics* 28, 78–88.

Wattayakorn, G., E. Wolanski and B. Kjerfve (1990). Mixing, trapping and outwelling in the Khlong Ngao mangrove swamp, Thailand. *Estuarine, Coastal and Shelf Science* 31, 5, 667–688.

Williams, D.McB. (1988). Significance of coastal resources to sailfish and black marlin in NE Australia: an ongoing research program. *Proceedings International Billfish Symposium II*, Kailua-Kona, Hawaii, 15 August, 1988.

Williams, D.McB. and S. English (1992). Distribution of fish larvae around a coral reef: direct detection of a meso-scale, multi-specific patch? *Continental Shelf Research* 12, 923–938.

Williams, D.McB., E. Wolanski and J. C. Andrews (1984). Transport mechanisms and the potential movement of planktonic larvae in the central region of the Great Barrier Reef. *Coral Reefs* 3, 229–236.

Willis, B. L. and J. K. Oliver (1990). Direct tracking of coral larvae: implications for dispersal studies of planktonic larvae in topographically complex environments. *Ophelia* 31, 145–162.

Wilson, P. R. (1985). Tidal studies in the One Tree Island lagoon. *Australian Journal of Marine and Freshwater Research* 36, 139–156.

Winant, C. D. and R. C. Beardsley (1978). A comparison of shallow currents induced by wind stress. *Journal of Physical Oceanography* 9, 218–220.

Wolanski, E. (1982). Low-level trade winds over the western Coral Sea. *Journal of Applied Meteorology* 21, 881–882.

Wolanski, E. (1983). Tides in the northern Great Barrier Reef continental shelf. *Journal of Geophysical Research* 88, 5953–5959.

Wolanski, E. (1986a). An evaporation-driven salinity maximum zone in Australian tropical estuaries. *Estuarine, Coastal and Shelf Science* 22, 415–424.

Wolanski, E. (1986b). Observations of wind-driven surface gravity waves offshore from the Great Barrier Reef. *Coral Reefs* 4, 213–219.

Wolanski, E. (1986c). Water circulation in a topographically complex environment. In: J. van de Kreeke (ed.), *Physics of Shallow Estuaries and Bays*. Springer-Verlag, Berlin, pp. 154–167.

Wolanski, E. (1986d). Island wakes and internal tides in stratified shelf waters. *Annales Geophysicae* 4, B4, 425–440.

Wolanski, E. (1986e). A simple analytical model of low-frequency wind-driven upwelling on a continental slope. *Journal of Physical Oceanography* 16, 1694–1702.

Wolanski, E. (1987). Some evidence for boundary mixing near coral reefs. *Limnology and Oceanography* 32, 3, 735–739.

Wolanski, E. (1992a). Hydrodynamics of tropical coastal marine systems. In: D. W. Connell and D. W. Hawker (eds.), *Pollution in Tropical Aquatic Systems*, CRC Press, Boca Raton, FL.

Wolanski, E. (1992b). Hydrodynamics of mangrove swamps and their coastal waters. *Hydrobiologia*, 247, 141–161.

Wolanski, E. (1993). Facts and numerical artifacts in modeling the dispersal of crown-of-thorn starfish larvae in the Great Barrier Reef. *Australian Journal of Marine and Freshwater Research*, 44, 3, 427–436.

Wolanski, E. and M. Jones (1981). Physical properties of Great Barrier Reef lagoon waters near Townsville. I. Effects of Burdekin River floods. *Australian Journal of Marine and Freshwater Research* 32, 305–319.

Wolanski, E. and B.. Ruddick (1981). Water circulation and shelf waves in the northern Great Barrier Reef lagoon. *Australian Journal of Marine and Freshwater Research* 32, 721–740.

Wolanski, E. and A. F. Bennett (1983). Continental shelf waves and their influence on the circulation around the Great Barrier Reef. *Australian Journal of Marine and Freshwater Research* 34, 23–47.

Wolanski, E. and D. van Senden (1983). Mixing of Burdekin River flood waters in the Great Barrier Reef. *Australian Journal of Marine and Freshwater Research* 34, 49–63.

Wolanski, E. and G. L. Pickard (1983). Upwelling by internal tides and Kelvin waves at the continental shelf break on the Great Barrier Reef. *Australian Journal of Marine Freshwater Research* 34, 65–80.

Wolanski, E. and R. E. Thomson (1984). Wind-driven circulation on the northern Great Barrier Reef continental shelf in summer. *Estuarine, Coastal and Shelf Science* 18, 271–289.

Wolanski, E. and G. L. Pickard (185). Long-term observations of currents on the central Great Barrier Reef continental shelf. *Coral Reefs* 4, 47–57.

Wolanski, E. and P. Ridd (1986). Tidal mixing and trapping in mangrove swamps. *Estuarine, Coastal and Shelf Science* 2, 759–771.

Wolanski, E. and W. H. Hamner (1988). Topographically controlled fronts in the ocean and their biological influence. *Science* 241, 177–181.

Wolanski, E. and P. Ridd (1990). Mixing and trapping in Australian tropical coastal waters. In: R. T. Cheng (ed.), *Residual Currents and Long-Term Transport.* Springer-Verlag, New York, pp. 165–183.

Wolanski, E. and B. King (1990). Flushing of Bowden Reef lagoon, Great Barrier Reef. *Estuarine, Coastal and Shelf Science* 31, 789–804.

Wolanski, E. and M. Eagle (1991). Oceanography and sediment transport in the Fly River estuary and Gulf of Papua. *Proceedings 10th Australasian Conference on Coastal and Ocean Engineering,* Auckland, 453–457.

Wolanski, E. and R. Gibbs (1992). Resuspension and clearing of dredge spoils after dredging, Cleveland Bay, Australia. *Water Environment Research* 64, 910–914.

Wolanski, E., M. Jones and W. T. Williams (1981). Physical properties of Great Barrier Reef lagoon waters near Townsville. II. Seasonal fluctuations. *Australian Journal of Marine and Freshwater Research* 32, 321–334.

Wolanski, E., J. Imberger and M. L. Heron (1984a). Island wakes in shallow coastal waters. *Journal of Geophysical Research* 89, 553–569.

Wolanski, E., G. L. Pickard and D. L. B. Jupp (1984b). River plumes, coral reefs and mixing in the Gulf of Papua and the northern Great Barrier Reef. *Estuarine, Coastal and Shelf Science* 18, 291–314.

Wolanski, E., E. Drew, K. M. Abel and J. O'Brien (1988a). Tidal jets, nutrient upwelling and their influence on the productivity of the alga Halimeda in the Ribbon Reefs, Great Barrier Reef. *Estuarine, Coastal and Shelf Science* 26, 169–201.

Wolanski, E., P. Ridd and M. Inoue (1988b). Currents through Torres Strait. *Journal of Physical Oceanography* 18, 1535–1545.

Wolanski, E., D. Burrage and B. King (1989). Trapping and dispersion of coral eggs around Bowden Reef, following mass coral spawning. *Continental Shelf Research* 9, 5, 479–496.

Wolanski, E., Y. Mazda, B. King and S. Gay (1990). Dynamics, flushing and trapping in Hinchinbrook Channel, a giant mangrove swamp, Australia. *Estuarine, Coastal and Shelf Science* 31, 555–579.

Wolanski, E., R. J. Gibbs, Y. Mazda, A. Mehta and B. King (1992a). The role of turbulence in the settling of mud flocs. *Journal of Coastal Research* 8, 1, 35–46.

Wolanski, E., R. Gibbs, P. Ridd and A. Mehta (1992b). Settling of ocean-dumped dredged material, Townsville, Australia. *Estuarine, Coastal and Shelf Science* 35, 473–490.

Wolanski, E., A. Norro and B. King (1992c). Fate of freshwater discharges in the Gulf of Papua, Papua New Guinea. Australian Institute of Marine Science Report to Ok Tedi Mining.

Wolanski, E., P. Ridd,, B. King and M. Trenorden (1992d). Fine sediment transport, Fly River estuary, Papua New Guinea. Australian Institute of Marine Science Report to Ok Tedi Mining Ltd.

Wolanski, E., B. Delesalle, V. Dufour, and A. Aubanal (1993). Modeling the fate of pollutants in the Tia-hurd lagoon, Moorea, French Polynesia. Proceedings 11th Australasian Conference on Coastal and Ocean Engineering, Townsville, 23–27 August 1993.

Yonge, C. M. (1940). The biology of reef-building corals. Science Report Great Barrier Reef Expedition, British Museum (Natural History), 1, 353–389.

Young, I. R. (1988). A parametric hurricane wave prediction model. *ASCE J. Waterway, Port, Coastal and Ocean Engineering* 114, 637–652.

Young, I. R. (1989). Wave transformation over coral reefs. *Journal of Geophysical Research* 94, C7, 9779–9789.

Young, I. R. and R. J. Sobey (1980). A predictive model of tropical cyclone wind-waves. *Proceedings 7th Australasian Conference on Hydraulics and Fluid Mechanics,* Brisbane, 812 August, 1980, 480–483.

Young, I. R. and R. J. Sobey (1981). Numerical prediction of tropical cyclone wind waves. Department of Civil and Systems Engineering, James Cook University, Technical Bulletin CS-20, 378 pp.

Young, I. and T. A. Hardy (1993). Measurement and modeling of tropical cyclone waves in the Great Barrier Reef. *Coral Reefs,* 12, 85–95.

Index

D

E

S

T

U

V

W

Y